1995

ENVIRONMENTALISM
AND
POLITICAL THEORY

SUNY Series in Environmental Public Policy
Lester Milbrath, editor

ENVIRONMENTALISM
AND
POLITICAL THEORY

Toward an Ecocentric Approach

Robyn Eckersley

State University of New York Press

Published by
State University of New York Press, Albany

For information, address State University of New York
Press, State University Plaza, Albany, N.Y. 12246

Production by Christine Lynch
Marketing by Fran Keneston

Library of Congress Cataloging-in-Publication Data

Eckersley, Robyn, 1958–
    Environmentalism and political theory : toward an ecocentric
approach / Robyn Eckersley.
        p.   cm. — (SUNY series in environmental public policy)
    Includes bibliographical references and index.
    ISBN 0–7914–1013–7 (alk. paper) . — ISBN 0–7914–1014–5 (pbk. :
alk. paper)
        1. Green movement. 2. Ecology. 3. Economic development-
-Environmental aspects. I. Title. II. Series.
JA75.8.E26   1992
320.5—dc20
                                                                        91–4059
                                                                        CIP

10 9 8 7 6 5 4 3 2 1

To Angus MacIntyre
1949–1991
fountain of hope and inspiration

Contents

Acknowledgments

When I began research for this book in 1985, Green political thought was a relatively new, relatively undeveloped, and reasonably manageable field of inquiry. By 1991 the literature on Green political ideas had expanded rapidly—a development that reflects the increasing international public concern over environmental issues and the growing international prominence of Green political parties during the latter half of the 1980s. I would not have been able to keep up to date with this burgeoning Green literature were it not for my Australian and international colleagues. I am also grateful to many of these same colleagues for their assistance, feedback, and support during the writing of this manuscript.

First and foremost, I would like to extend a very special thank you to Warwick Fox for inspiration, many long and stimulating discussions, detailed and invaluable critical feedback, general encouragement, and whole-hearted support in more ways than I could possibly enumerate. I am also indebted to Peter Hay for his feedback, encouragement, and proofreading and to John Todd and the rest of the staff and students at the Centre for Environmental Studies for providing such a friendly and supportive environment in which to research.

A special thanks must also go to Clay Morgan and Lester Milbrath of SUNY Press for their editorial support and encouragement and to Alan Drengson, Robert Paehlke, John Rensenbrink, and Peter Wenz for their insightful feedback and assistance as SUNY readers.

I would also like to extend a thank you to my other colleagues and correspondents, particularly David Abram, Don Alexander, Murray Bookchin, John Clark, Nigel Clark, Bill Devall, Riley Dunlap, Paul and Anne Ehrlich, Patsy Hallen, the late Angus MacIntyre, Bill McCormick, Andrew McLaughlin, Freya Mathews, David Orton, Sara Parkin, Jonathon Porritt, Lorna Salzman, George Sessions, Brian Tokar, Shann Turnbull, John Young, and Michael Zimmerman for a stimulating exchange of ideas and/or for providing me with books, papers, inspiration, information, and feedback.

I am also grateful to the Australian federal government for an Australian Research Council Postdoctoral Fellowship, which has enabled me to streamline, revise, and rework my doctoral dissertation into this book.

Finally, I would like to express my sincere gratitude to my parents for their constant love and support for as long as I can remember.

Chapter 5 is a revision and further development of my article "Habermas and Green Political Thought: Two Roads Diverging" which appeared in *Theory and Society* 19 (1990): 739–76.

Introduction

> Our ecologic crisis is the product of an
> emerging, entirely novel, democratic
> culture. The issue is whether a
> democratized world can survive its own
> implications. Presumably we cannot
> unless we rethink our axioms.
>
> Lynn White, Jr.[1]

Nowadays, the notion of the intrinsic value or dignity of the individual—and the democratic culture to which it has given rise—is so obvious and compelling to the Western imagination that very few political thinkers find it necessary to explore its content, ambit, and basis. As Steven Lukes points out, the notion of human dignity or respect for all persons is one of the most basic ideas that has come to pervade modern social and political thought in the West. For example, it is now

> enshrined in the American Declaration of Independence, in the Declaration of the Rights of Man and in the Universal Declaration of Human Rights adopted by the General Assembly of the United Nations in 1948, which began by declaring its "recognition of the inherent dignity and of the equal and inalienable rights of all members of the human family."[2]

This noble humanist ideal of the dignity of the person has a long and complex pedigree in Western thought and may be traced to both classical Greek and Christian sources. Moreover, as Blackham points out:

> The peak periods of "humanisms," namely, the Greek Enlightenment, the Renaissance, the European Enlightenment and its prolongation into various movements of the nineteenth century…were formative periods that transformed a dominant part of the original Europe of the Church into modern secular industrial democracies.[3]

1

Yet the ecological crisis has exposed the hitherto unexamined flip side of our Western humanist heritage. In the face of accelerated environmental degradation and species extinction, environmental philosophers are now asking: are we humans the *only* beings of value in the world? Does the world exist only for *our* benefit?

It is important to emphasize that this environmental philosophical inquiry has not been directed to the notion of human dignity or intrinsic value per se. Rather, it has been directed to the fact *that intrinsic value is generally taken to reside exclusively, or at least preeminently, in humans*—a belief that environmental philosophers argue has resulted in the systematic favoring of human interests over the interests of the nonhuman world. As David Ehrenfeld observes, although humanism "has its nobler parts" and therefore ought not to be rejected,

> we have been too gentle and uncritical of it in the past, and it has grown ugly and dangerous. Humanism itself, like the rest of our existence, must now be protected against its own excesses. Fortunately, there are humane alternatives to the arrogance of humanism.[4]

The questions raised by environmental philosophers have exposed a number of significant blind spots in modern political theory. This book seeks to respond to the environmental philosophical challenge by accepting that we must indeed widen the ambit of political discussion to include the question of our relationship to, and impact upon, the nonhuman world. Questions of this kind have rarely been aired explicitly by political theorists, let alone given any prominence. In the main, political theorists have confined their attention to social questions and affairs of state, that is, to inter-human relations rather than human-nonhuman relations. Of course, every political worldview incorporates certain assumptions and values bearing on the relationship of humans to the rest of nature.[5] However, these assumptions and values have usually been considered relevant only insofar as they help us to see what is special about humans vis-à-vis the nonhuman world. The latter has attracted attention only insofar as it is instrumentally valuable to human actors, whether as a material resource or as some other means to human self-realization. In short, the nonhuman world is generally accorded the status of a background or stage upon which the human drama unfolds. Invariably, these assumptions concerning the relationship between the human and nonhuman world are opaque and uncontroversial since they usually form part of the stock assumptions of the age. Yet it is surprising how the fragments of certain cultural and political assumptions characteristic of one particular age can persist in subsequent times even when the foundations upon which such assumptions originally rested have been supplanted or seriously challenged by, say, new scientific discoveries or new philosophical investigations. As Lynn White has observed:

Despite Copernicus, all the cosmos rotates around our little globe. Despite Darwin, we are *not*, in our hearts, part of the natural process. We are superior to nature, contemptuous of it, willing to use it for our slightest whim.[6]

More generally, Alvin Gouldner has noted how "old background assumptions"—the "inherited intellectual 'capital' with which a theorist is endowed"—may come to operate in new conditions and act as boundaries that confine and inhibit the further development of a particular theoretical enterprise.[7] It is part of the burden of this inquiry to uncover and challenge some of the stock assumptions of modern political theory and to show how they have inhibited a deeper appreciation of our environmental ills and perpetuated the very processes of environmental destruction.

This critique of anthropocentrism—or human centeredness—will prepare the way for the outline of a more encompassing, nonanthropocentric political framework with which to approach social and ecological problems. The outline of this alternative framework, which rests on an ecocentric (i.e., ecology-centered) perspective, will emerge in the course of a critical evaluation of the principal new ideas that make up the melting pot of Green political thought.

Whether the ecocentric approach that I defend is understood as breaking new ground or simply reworking familiar political themes ultimately depends on the criterion that is used to distinguish one political theory from another. From the perspective of the conventional left-right spectrum of political thought, the ecocentric political theory defended in this book might be seen as just another moderately left-of-center permutation of familiar political themes. Viewed in the light of the anthropocentric and cornucopian assumptions of post-Enlightenment political thought, however, such an ecocentric approach may be seen as a genuinely new constellation of political ideas.

The book is divided into two parts. Part 1 is mainly concerned to explain and defend the ecophilosophical foundations of ecocentrism, while part 2 is mainly concerned to examine a range of specific Green political theories and, through a dialogue with these theories, find out how a general ecocentric perspective might be fleshed out in a political and economic direction.

In chapter 1, I characterize Green political theory as being concerned to reconcile the themes of participatory democracy and human survival through a more encompassing theme of ecological emancipation. I then divide Green political theory into an anthropocentric stream and an ecocentric stream and, in chapters 2 and 3, articulate and defend an ecocentric perspective in the course of a discussion of some of the central debates and arguments that have been advanced in the new but rapidly growing field of environmental philosophy. The discussion in chapter 2 includes an exploration of the major currents of modern environmentalism (resource conservation, human welfare ecology, preservationism, animal liberation, and ecocentrism). In chapter 3, I also discuss three distinctive ecocentric philosophical approaches, namely, autopoietic intrinsic value theory, deep ecology (alternatively referred to as transpersonal ecology), and ecofeminism.

In part 2, I turn to an examination of the major currents of Green political thought that are currently vying for ascendancy. My main concerns here are to (i) show where each current of Green political thought sits on the anthropocentric/ecocentric spectrum; (ii) assess the internal theoretical coherence of each synthesis and draw out the political priorities and institutional reforms that flow from each approach; and (iii) suggest, where appropriate, alternative institutional arrangements that might be more consistent with an ecocentric perspective. These Green political currents are identified under the broad generic names of eco-Marxism, which includes both orthodox and humanist eco-Marxism (chapter 4) as well as the Critical Theory of the Frankfurt School (chapter 5), ecosocialism (chapter 6), and ecoanarchism (chapter 7). Liberal and conservative responses to the ecological crisis do not qualify as Green political theories according to my classification schema and are dealt with summarily in chapter 1. Although the ecologically reworked version of orthodox Marxism also fails to qualify as a Green political theory, I have included an extended discussion of this theory in chapter 4 in order to provide a benchmark by which we may compare how far other ecologically oriented socialist theorists have moved away from many of the problematic tenets of orthodox Marxism.

I conclude that no single current of Green political thought examined in part 2 is capable of standing alone as an adequate ecocentric political theory. In the concluding chapter, I draw together the basic features of a comprehensive and feasible ecocentric political theory that has emerged from the critical dialogue with existing currents of Green political thought.

PART I

Staking out the Green Terrain

CHAPTER 1

The Development of Modern Ecopolitical Thought: From Participation and Survival to Emancipation

INTRODUCTION

The environmental crisis and popular environmental concern have prompted a considerable transformation in Western politics over the last three decades. This transformation has culminated in the development of new political cleavages, the formation of Green political parties, and the revision of old political platforms by existing parties.[1] This, in turn, has generated new areas of political study and analysis as journalists, academics, and other observers seek to come to terms with these new political developments. Whatever the outcome of this realignment in Western politics, the intractable nature of environmental problems will ensure that environmental politics (or what I shall refer to as "ecopolitics") is here to stay.

How can we make sense of these new political developments? If we look back over the ecopolitical literature of the last three decades, it is possible to discern three major ecopolitical preoccupations that can be encapsulated in the themes of participation, survival, and emancipation. To some extent, these three themes may be seen as roughly characterizing the general ecopolitical preoccupations of the last three decades respectively, although this temporal association is a loose one only and should not be pressed too far (i.e., "later" ecopolitical themes are discernible in earlier periods just as "earlier" themes are discernible in subsequent periods). Indeed, the last three decades have seen a general broadening of ecopolitical dialogue as a result of the gradual interpenetration of these themes or phases of inquiry. That is, the participatory, survivalist, and emancipatory phases may be seen as representing the thesis, antithesis, and higher synthesis respectively in the ecopolitical dialogue of the last three decades.

The emergence of a general Green philosophy and Green political parties in the late 1970s and 1980s may be seen as representing this third emancipatory moment. That is, although Green political thought sometimes draws on the New Left participatory thinking of the 1960s and some aspects of the survivalist or

7

so-called doomsday environmental literature of the early 1970s, it nonetheless represents a new, ecologically inspired political orientation that has raised new political issues and called into question old political responses. Of course, the label *Green* is an extraordinarily elastic one that has been applied to, or appropriated by, all manner of environmental and political positions over the past decade. However, from the point of view of participants in the Green movement and in Green political parties, the word *Green* represents a distinctive body of ideas and a new political force. The developmental, tripartite characterization of ecopolitical thought presented in this chapter is offered as one way in which we may locate and distinguish the broad contours of a general Green (i.e., emancipatory) political perspective from other ecopolitical perspectives. Of course, within this broad Green ballpark there are many different and more subtle interpretations of the theory and practice of Green politics. Indeed, we shall see that Green politics has its own internal spectrum of debate, with its own competing "political wings." What is noteworthy, however, is that the most important of these wings are not the familiar left and right wings of conventional politics, although such divisions can be found. Rather, the most significant internal difference is to be found between what I shall call *anthropocentric Greens* and *ecocentric Greens*. As we shall see, this new environmental cleavage can be used to shed very new light on very old political debates.

THE ENVIRONMENTAL PROBLEMATIC AS A CRISIS OF PARTICIPATION

The 1960s marked the beginning of widespread public concern over environmental degradation in the developed countries of the West.[2] However, it took roughly a decade of persistent political agitation over such matters as pesticides, nuclear power plants, toxic waste dumps, large scale industrial developments, and pollution before an "environmental crisis" was officially recognized as a matter of local, national, and international concern.[3] The first Earth Day celebrations in 1970, the emergence of a panoply of new environmental laws in Western countries in the late 1960s and early 1970s, the development of interdisciplinary environmental studies programs in higher education institutes, and the United Nations Conference on the Human Environment at Stockholm in 1972 all represent significant landmarks of national and international recognition of environmental problems.

Yet much of this official recognition, such as new environmental legislation, also helped to define and contain environmental problems as essentially matters of poor planning rather than as indicators that the optimistic and cornucopian assumptions of the post–World War II growth consensus might need to be revised. In particular, the notion that there might be ecological limits to economic growth that could not be overcome by human technological ingenuity and better planning was not seriously entertained until after the much publicized "limits to growth" debate of the early 1970s. As John Rodman observed in the context of the United States, environmental problems were originally perceived in the 1960s as a "crisis of participation" whereby excluded groups sought to ensure a more equitable distribution of environmental "goods" (e.g., urban amenity) and "bads" (e.g., pollution).[4] This is not surprising given that the early wave of environmental activism was generally seen as but a facet of the civil rights movement in its concern for more grassroots democratic participation in societal decision making, in this case, land and resource usage. The growth in public concern over environmental problems was thus widely interpreted as being only, or at least primarily, concerned with participatory and distributional issues, that is, issues concerning "who decides" and "who gets what, when, and how." The upshot was that by the 1970s environmental problems were, as Rodman has put it,

> domesticated by mainstream political science, reduced to the study of pollution control policy and environmental interest groups, and eventually absorbed within the framework of "the policy process" and the "politics of getting."[5]

This kind of characterization of the problem was widely shared by both policy makers and political theorists. This is not to say that new critiques, sensibilities, and theoretical paths did not emerge in the 1960s and early 1970s. Rachel Carson's *Silent Spring,* Murray Bookchin's *Our Synthetic Environment,* and, to a lesser extent, Charles Reich's prophetically titled *The Greening of America* represent three important landmarks in the emergence of a new sensibility that celebrated the living world and was deeply critical of dominant Western attitudes toward the nonhuman world.[6] These contributions were, however, exceptions. By and large, there were few major theoretical innovations in social and political thought in the 1960s that arose specifically from a consideration of the environmental crisis. This tendency to treat environmental protest as an aspect of the wider pursuit of distributive justice and democratic planning was especially marked among socialist, social democratic, and liberal welfare theorists—a tendency that has continued through the 1980s. Perhaps the exemplar of this kind of social democratic analysis is Hugh Stretton's award winning book *Capitalism, Socialism and the Environment,* which opens with the unequivocal declaration that:

> This book is about the distribution of environmental goods: the shares that go to rich and poor in the developed democracies of Europe, North America, Japan and Australasia.[7]

Although I must hasten to add that distributional questions remain crucial questions in any ecopolitical inquiry, to circumscribe the problem in this way can nonetheless serve to reinforce rather than challenge the prevailing view that the environment is simply a human resource (albeit a resource to be utilized more efficiently and equitably).

By virtue of the radical democratic and participatory nature of environmental protest in the 1960s, many political commentators tended to regard it as an adjunct of the New Left. Yet even this association was soon to come under challenge as a rearguard action developed against environmentalism by labor, socialist, and liberal welfare activists and theorists. The discovery of the socially regressive consequences of some environmental reforms (e.g., the costs of pollution abatement being passed on as higher prices; unemployment resulting from the closing down of polluting industries) soon gave rise to the now familiar accusation that environmental protest is an elitist, middle-class phenomenon that threatens the hard won material gains and jobs of the working class.[8] Such social conflicts provide a significant indication of the gradual realignment of political cleavages that has been taking place in the West between the so-called New Class (or New Middle Class) that furnishes the core activists of the environmental movement and the two traditional classes of industrial society, namely, the owners/controllers of capital and the working class.[9] The growing tension that developed between the demand for environmental reform, on the one hand, and redistributive justice and economic security, on the other hand, has remained an enduring and vexed issue in ecopolitical discussion.

The 1960s and early 1970s were also a time of theoretical stocktaking and revision for socialist theory—a revision spearheaded by the rise of the New Left. In particular, Herbert Marcuse's *One Dimensional Man* and the essays collected in Jürgen Habermas's *Toward a Rational Society* played an influential role in tracing many of the problems of industrial society—including its environmental dislocations—to the dominance of instrumental or technocratic rationality.[10] This contributed to the widening of the New Left's agenda to include questions of life-style, technology, and the exploitation of nature. The ready absorption of these ideas by the counterculture and "back-to-nature" movements of the 1960s was defended eloquently by political theorists and cultural historians such as Murray Bookchin, Theodore Roszak, and Charles Reich.[11] Many of the issues raised by these writers, such as the importance of consciousness change and alternative worldviews, remain significant currents in modern emancipatory/Green theorizing.

Yet, with the exception of the work of Roszak and, to some extent, Marcuse and Bookchin (whose contributions are explored in chapters 5 and 7), none of these early theoretical developments mounted a serious challenge to anthropocentrism or argued for a new humility and compassion in our dealings with the nonhuman world. Rather, the growing concern for environmental quality was incorporated into the New Left's agenda for greater individual and community autonomy and control. After all, the overriding revolutionary goal of the New Left, as George Katsiaficas succinctly describes it in his comprehensive international study, was "the decentralization and self-management of power and resources."[12]

To most New Left thinkers, then, questions concerning humanity's power vis-à-vis the rest of nature were dealt with in terms of who exercised such power and on whose behalf. Of course, these were (and still are) crucial questions, as I have already noted, yet they remained embedded in an essentially anthropocentric framework and were firmly wedded to the "participatory" ethos of the times. As we shall see in chapter 5, even the innovative attacks on the ideology of "scientism" and instrumental rationality waged by Critical Theorists such as Marcuse and Habermas only partially transcended this framework (enough, however, for me to identify these theorists as emancipatory theorists, albeit in the anthropocentric rather than ecocentric stream). Their overriding concern was to open up improved channels of political communication in order to facilitate the achievement of a democratic consensus that would direct the development and use of technology toward more human liberatory ends. This was also the major thrust of William Leiss's critique of "the domination of nature."[13] Although these critiques were innovating and provocative and remain important contributions to emancipatory thought, their overriding objective was the liberation of "inner" rather than "outer" nature (i.e., human instincts or human communication rather than the nonhuman community). As we shall see in chapter 5, Critical Theory's objection to the domination of nature ultimately comes to rest on the human-centered argument that it leads to the domination of *people*.

This brief characterization and critique of the participatory theme in ecopolitical thought should not be interpreted as a rejection of the contribution of the New Left. Questions concerning citizen participation, self-management, and distributive justice remain central issues in emancipatory thought. These themes are reflected, for example, in two of the so-called four pillars upon which the platforms of many Green parties rest, namely, grassroots democracy and social justice.[14] Yet these themes are merely a necessary—as distinct from a sufficient—condition for a proper characterization of emancipatory thought. That is, while these themes are essential aspects of Green political thought, they do not *distinguish* Green thought from other ecopolitical perspectives. The "discovery" of "ecological interconnectedness"—which was brought to public attention in the early 1960s with the publication of Rachel Carson's *Silent Spring* but did not gather momentum until the late 1970s and early 1980s—was to set in train significant theoretical innovations, the *political* repercussions of which are only beginning to be worked out in any degree of detail. As we shall see, the most significant of these has been the attempt by emancipatory theorists to revise and incorporate the principles of individual and community autonomy into a broader, ecological framework.

THE ENVIRONMENTAL PROBLEMATIC AS A CRISIS OF SURVIVAL

The "crisis of survival" theme in ecopolitics rose to prominence in the early 1970s following the publication of the Club of Rome's *The Limits to Growth* and *The Ecologist* magazine's *Blueprint for Survival*.[15] Although evidence of widespread environmental deterioration had been steadily accumulating since the 1950s, the sensational and widely publicized findings of these two reports posed a considerable challenge to the sanguine belief that we could all continue with business and politics as usual.[16] The mounting evidence of environmental degradation stemming from the exponential growth in resource consumption and human population was shown to pose very real threats to the earth's biological support systems. Although there were important differences between the two reports they shared the same general message. That is, the environmental crisis amounted to much more than a crisis of participation: what was at stake was the very survival of humanity.[17] The metaphor of our planet as spaceship Earth—which had become popular following the circulation of images of the "Whole Earth" taken from outer space by NASA—was widely employed to emphasize a new appreciation of the fragility and finiteness of the Earth as an "oasis in the desert of infinite space."[18] This marked the emergence of a deeper appreciation of the *global* dimensions of environmental degradation and the *common fate* of humanity. However, some of the ecopolitical solutions offered in the wake of this new awareness of global environmental degradation and resource scarcity (such as Garrett Hardin's "life-boat ethics") did not prove to be particularly "brotherly."

Not surprisingly, the dire projections of *The Limits to Growth* and *Blueprint for Survival* (which carried the endorsement of many eminent British scientists) had a significant impact on the world's media and prompted calls for a swift and multifaceted response from national governments. (The ensuing debate was intensified by the 1973–74 oil crisis, which came as a timely reminder of the heavy oil dependence and hence vulnerability of industrialized countries.) Indeed, *The Ecologist*'s detailed solution outlined in *Blueprint for Survival* provided the impetus for the formation in 1973 of Europe's first Green party, the British People's Party (which later became the Ecology Party in 1975 and the British Green Party in September 1985). This party adopted *The Ecologist*'s radical "blueprint" as its basic theoretical statement.[19] *Blueprint for Survival* has proved to be a landmark publication in Green politics in foreshadowing many of the goals and policies that are found in the platforms of the various Green parties that formed in the late 1970s and 1980s.

In concentrating mainly on the physical limits to growth, however, the MIT study commissioned by the Club of Rome spawned a plethora of counterarguments to the effect that the problems were susceptible to "technological fix" and pricing solutions that would alleviate the negative ecological externalities of economic growth without the need for any fundamental changes in political values or the pattern and scale of economic activity. Moreover, the particular projections of the MIT team were criticized for containing methodological flaws and resting on unduly pessimistic assumptions.[20]

Yet the methodological problems that have been discovered in *The Limits to Growth* have not, by and large, seriously detracted from its essential message. The Club of Rome's 1974 updated survey (prepared in response to criticisms of its 1972 report) concluded, in a slightly more optimistic tone, that growth was possible, provided it was ecologically benign:

> For the first time in man's life on earth, he is being asked to refrain from doing what he can do; he is being asked to restrain his economic and technical advancement, or at least to direct it differently from before; he is being asked by all future generations of the earth to share his good fortune with the unfortunate—not in a spirit of charity, but in a spirit of survival.[21]

Indeed, many of those who have been most critical of this body of so-called doomsday literature have acknowledged that the crisis is real and that far-reaching changes in both our values and institutions are required if ecological and social catastrophe is to be averted.[22] Moreover, the basic message of *The Limits to Growth* and *Blueprint for Survival* has been reinforced by later, more refined studies of global trends in population growth, resource consumption, and ecological deterioration. For example, the major study of the world's environmental problems commissioned by President Carter in *The Global 2000 Report to the President of the U.S.* summarized its findings as follows:

> If present trends continue, the world in 2000 will be more crowded, more polluted, less stable ecologically, and more vulnerable to disruption than the world we live in now. Serious stresses involving population, resources, and environment are clearly visible ahead. Despite greater material output, the world's people will be poorer in many ways than they are today.[23]

The annual *State of the World* reports, published by the Washington based Worldwatch Institute, and the recent Brundtland Report (*Our Common Future*) have continued to reinforce this message.[24]

Not surprisingly, many of the ecopolitical publications that appeared in the climate of the early 1970s—especially those that appeared in the immediate aftermath of the "limits to growth" debate—shared an overriding preoccupation with human survival, a sense of great urgency, a new, practical and empirical frame of mind, and a preparedness to call for tighter governmental controls.[25] Gone were the heady New Left calls for freedom, citizen participation, and the "good life." In their stead came sober discussions of resource rationing, increasing government intervention, centralization, and population control. The new message, expressed eloquently by Robert Heilbroner in the closing pages of *An Inquiry into the Human Prospect* (a landmark survivalist publication that typified the mood and temper of the period), was that the individualistic Promethean spirit must give way to the example of Atlas—the spirit of fortitude, resolutely bearing whatever burdens were necessary to sustain life.[26] Appropriately, the cover of Heilbroner's book bears a picture of a doleful Atlas, stoically bearing the load of the Earth on his shoulders.

As early as 1968, Garrett Hardin set the tone of this phase of the discussion in his influential essay "The Tragedy of the Commons" with his warning that freedom in the unregulated commons brings ruin to all.[27] Hardin's well-known parable of the medieval herdsmen overstocking the commons vividly demonstrates the tragic dynamic that arises when people are motivated by an economic "rationality" that has as its sole objective the maximization of individual gain in the short term. Hardin has argued that when people act according to such an economic rationality they will inevitably despoil the commons, even when they have full knowledge of the mounting public cost that the pursuit of private gain will bring.[28] Hardin's answer to the tragedy— "mutual coercion, mutually agreed upon by the majority of the people affected"—marked this survivalist school as one whose overriding preoccupation was to find the means of warding off disaster and discover a minimally acceptable way of life rather than search for the "good life."[29]

Hardin did not, however, extend his ecosocial contract theory (which rested on *mutual agreement* by the *majority* of the people affected) to the global population problem. His notorious neo-Malthusian "life-boat ethic," which argued against a more equitable distribution of the world's resources on the grounds that we would *all* "go under," has been widely condemned for protecting the advantages of the affluent and pronouncing a death sentence for the poor.[30] As Richard Barnet has argued:

> The specter of the hungry mob supports Hobbesian politics, a world of struggle over inadequate resources that cries out for Leviathan, the authoritarian state that can keep minimal order. The Malthusian fantasy offers an alternative to the Leviathan state. There is no need for a civil authority to regulate scarce goods, because Nature, cruel only to be kind, periodically thins the surplus population by famine.[31]

As we have seen, the general preoccupation with survival also stamped Heilbroner's somber inquiry, which opened with the searching question: "Is there hope for humanity?" After exploring world demographic trends in the context of the persistent threat of nuclear war and the escalation of environmental degradation, Heilbroner reached a reluctant and pessimistic conclusion. Given "human nature" (which Heilbroner saw as fundamentally selfish), our only hope for survival lay in our obedient rallying behind a centralized, authoritarian nation—the only institutional form that Heilbroner saw as capable of extracting the necessary sacrifices, regulating distribution, and redirecting agriculture and industry along ecologically sustainable lines.

Since Heilbroner's major concern was the fundamental issue of human survival, he did not address (and, at the time, would probably have thought it a luxury to consider) the question of how to preserve and foster the more agreeable aspects of human nature, at least during the convulsive period of transition. Faced with the urgency of the interrelated crises confronting humankind (particularly the environmental crisis), Heilbroner adopted an empirical frame of mind, focusing on how people are likely to behave rather than on what people might eventually become. In this context, he insisted that we cannot afford to ignore obdurate human characteristics and build a future on unrealistic beliefs.[32] In Heilbroner's assessment, people will not willingly acquiesce in giving up a way of life, particularly where it entails the enjoyment of relative privileges. It is this premise that set the tone of Heilbroner's entire analysis.

It deserves mention, however, that although Heilbroner saw centrally planned, authoritarian states as the necessary transitional scenario, it is clear that this is not what he would personally wish for. Rather, he expressed a preference for "a diminution in scale, a reduction in the size of the human community from the dangerous level of immense nation states toward the 'polis' that defined the appropriate reach of political power for the ancient Greeks."[33] In Heilbroner's view, however, this vision (which is the one generally promoted in *Blueprint for Survival*) was highly improbable in the short and immediate term.

Heilbroner's political conclusion—that external constraints on human behavior are essential to make possible the transition from a growth oriented to a steady state society—has also been endorsed to a large extent by William Ophuls.[34] Like Heilbroner, Ophuls also admits his preference for a smaller scaled, face-to-face democracy of the Greek city state or Jeffersonian type, which he sees as the most appropriate vehicle for the pursuit of "the good life," but he considers that "reforming a 'corrupt people' is a Herculean task" (recall Heilbroner's Atlas!). In Ophuls' view, we are ultimately confronted with a limited choice between "Leviathan or oblivion."[35] Although Ophuls has since moderated his position by placing a greater emphasis on the need for self restraint than on the need for external coercion, he continues to maintain that the latter must be resorted to if calls for the former are unsuccessful.[36]

Ophuls and Heilbroner may be seen as offering more interventionist variants of Hardin's call to "legislate temperance" by "mutual coercion, mutually agreed upon" in order to mitigate the ecologically and socially destructive rationality that characterizes human behavior in the unmanaged commons. Heilbroner's and Ophul's fellow Americans are seen as sharing the same characteristics as Hardin's herdsmen—"selfish hedonists rationally seeking private gain." They therefore have much in common with the model of the self-interested human who roamed in Hobbes's and Locke's state of nature insofar as they are seen as being in perpetual (Hobbes) or intermittent (Locke) conflict with the interests of the larger social and natural community to which they belong. In such a context, salvation can only come from the surrender of a considerable degree of individual liberty to a central authority. Indeed, Ophuls has frequent recourse to the social contract theories of the seventeenth and eighteenth centuries, suggesting that the constitutional limits of the central authority of the future might be struck in accordance with a new "ecological contract" that would (hopefully) be based on prudent self-restraint and seek harmony not only among humans but also between humans and the rest of nature.[37] However, unlike the social contract of Locke (which was based on cornucopian assumptions), the ecological contract would be based on the Hobbesian premise of scarcity and would therefore require an all powerful Leviathan, not just a limited government.[38] That is, if certain freedoms were not voluntarily surrendered by citizens, then restrictions would have to be imposed externally by a sovereign power.[39]

The authoritarian solutions proffered by Heilbroner and Ophuls and the life-boat ethics of Hardin have prompted a number of critics to ask just what is to be sacrificed in the name of human survival and to ponder whether perhaps the price might be too high.[40] In particular, the dire analyses of this survivalist school have been widely criticized (particularly, but not only, by socialist theorists) for displaying an insensitivity to old conflicts such as national rivalry and the gap between rich and poor. As Andrew Feenberg has observed, this insensitivity

> leads to a politics of despair that would freeze the current relations of force in the world—and with them the injustices they sustain—as a condition for solving the issue of survival.[41]

Similarly, Enzensberger has criticized those who employ the "brotherly" rhetoric of spaceship Earth for conveniently overlooking the difference between "the bridge and the engine room."[42] Others critics, reasserting the participatory theme, have argued that it is the very *erosion* of liberal democracy that has enabled powerful elites to pursue, with the backing of the State, environmentally destructive growth.[43] What is needed is *more* rather than less participation in government; the survivalists, according to this view, have seriously overestimated the capabilities of centralized institutions and underestimated the capabilities of decentralized, democratic political institutions to respond to the crisis.[44]

While agreeing with the need for more participation, some political theorists have expressed more deep-seated reservations about the capacity of liberal democracy to meet the ecological challenge. As Susan Leeson has put it:

> if authoritarianism is the response to the inability of popular governments to impose the limits required to avoid ecological disaster,

such a response merely reflects the crisis to which modern political philosophy and liberalism have led; it is not itself a solution.[45]

What is needed, these critics argue, is a fundamental reexamination of the basic axioms of liberalism such as possessive individualism, private property, limited government, and market freedom. According to Leeson:

> it was the unleashing of the passion for material abundance, legitimized by Hobbesian natural right, amplified by Locke, combined with the rejection of the classical commitment to reason and proper limits that caused the ecological crisis.[46]

It was this kind of ecological critique of liberalism that led many ecopolitical theorists to turn to the broad socialist tradition as an alternative. Yet, as we shall shortly see, other ecopolitical theorists found many of the ecologically problematic assumptions of liberalism to be also embedded in the socialist tradition.[47] From this important dialogue between survivalists and their critics there emerged the highly contested question: is socialism ecologically salvageable or must we look elsewhere, that is, beyond liberalism *and* socialism, for ecopolitical enlightenment?

Despite the widespread criticism of the authoritarian response to the deepening ecological crisis, it would be wrong to dismiss the survivalists' contribution out of hand. First, they have done much to draw attention to the seriousness of the ecological crisis and have challenged the widespread complacency concerning the ability of existing political values and institutions to respond to the crisis. Second, the controversial nature of the authoritarian solutions that surfaced in the wake of the "limits to growth" debate has encouraged the search for more deepseated cultural transformations along with alternative, nonauthoritarian institutions that would foster a more cooperative and democratic response to the environmental crisis. In this respect, the above authoritarian scenarios have become sobering reminders of what can and might happen if too little remedial action is taken, or if it comes too late. These scenarios have thus served as a useful foil for later democratically and ecologically oriented theorists who have sought to develop an alternative solution to the environmental crisis that incorporates yet revises and transcends the general participatory ethos of the 1960s, which had been largely premised on now discredited cornucopian assumptions.

THE ENVIRONMENTAL PROBLEMATIC AS A CRISIS OF CULTURE AND CHARACTER AND AS AN OPPORTUNITY FOR EMANCIPATION[48]

Many of those who were critical of the survivalist school responded by extending ecopolitical debate beyond the realm of the *physical* limits to growth to the point of questioning the very notion of material progress and lamenting the social and psychological costs associated with the dominance of instrumental rationality. Included among these costs were alienation, loss of meaning, the coexistence of extreme wealth and extreme poverty, welfare dependence, dislocation of tribal cultures, and the growth of an international urban monoculture with a concomitant reduction in cultural diversity.[49] For those who took this step, the sanguine reliance on future "technological fixes" and better planning—seen by many other critics of survivalism as the definitive rejoinder to the "limits to growth" projections—was increasingly recognized as part of the problem rather than the solution. By the late 1970s and early 1980s, a growing number of ecopolitical thinkers were pointing to the new cultural opportunities that lay in what had hitherto been pessimistically approached by the survivalists as a dire crisis with a limited range of options. In short, this new breed of ecopolitical theorists began to draw out what they saw as the emancipatory potential that they believed was latent within the ecological critique of industrialism. Moreover, this new project entailed much more than a simple reassertion of the modern emancipatory ideal of human autonomy or self-determination. It also called for a reevaluation of the foundations of, and the conditions for, human autonomy or self-determination in Western political thought.

The general tenor of this third, emancipatory phase of ecopolitical inquiry may be best introduced in the voices of some of its leading contributors. As William Leiss has explained:

> No elaborate argument should be necessary to establish that there are some limits to economic and population growth. But everything depends upon whether we regard such limits as a bitter disappointment or as a welcome opportunity to turn from quantitative to qualitative improvement in the course of creating a conserver society.[50]

John Rodman has sounded a similar theme in pointing out:

> to the extent that limits are perceived as external to us, they may have to be imposed on us by authoritarian governments; whereas the more they are perceived as arising from within personal and social experience—e.g., in the form of frustration resulting from the "limits to consumption"..., then the more the "limits to [industrial] growth" emerge "naturally," and the appropriate role for government appears, which is not to repress growth, but to stop forcing it...and to facilitate the transition to the steady state.[51]

As early as 1965 Murray Bookchin argued, in a prophetic and pioneering essay entitled "Ecology and Revolutionary Thought," that the insights of ecology offered a critique of society "on a scale that the most radical systems of political economy have failed to attain."[52] Since that time Bookchin has maintained the argument that the cultivation of an ecological society, resting on the principles of social ecology, will serve to *expand* rather than narrow the realm of freedom or self-directedness in first (i.e., nonhuman) and second (i.e., human) nature.

Theodore Roszak, another pioneer of this emancipatory approach to ecopolitics, has pointed to what he sees as the "vital reciprocity" between person and planet:

> My purpose is to suggest that the environmental anguish of the Earth has entered our lives as a radical transformation of human identity. The needs of the planet and the needs of the person have become one, and together they have begun to act upon the central institutions of our society with a force that is profoundly subversive, but which carries the promise of cultural renewal.[53]

Rudolf Bahro, in a somewhat ironic tone, has signalled his indebtedness to the environmental crisis because it has forced us to reexamine the question of emancipation in fresh terms. According to Bahro, if the Earth were infinite and if there were no problems of energy shortages and resource depletion, we would continue to believe (falsely, in Bahro's view) that the road to freedom lay in material expansion.[54] Bahro has argued that the environmental crisis, which he has claimed to be the "quintessential crisis of capitalism," has forced us to reexamine not only the psychological costs of the competitive and expansionary ethos of our materialist culture but also our imperialist attitude toward other species.

In a similar vein, Christopher Stone, in his eloquent defence of the "rights" of nonhuman beings, has regarded the environmental crisis as offering an opportunity for metaphysical reconstruction and moral development. In voicing the approach taken by a growing number of ecophilosophers, Stone has argued that:

> whether we will be able to bring about the requisite institutional and population growth changes depends in part upon effecting a radical shift in our feelings about "our" place in the rest of Nature.
>
> A radical new conception of man's relationship to the rest of nature would not only be a step towards solving the material planetary problems; there are strong reasons for such a changed consciousness from the point of making us far better humans.[55]

Pursuing this same theme, Bill Devall and George Sessions have argued for the cultivation of new "character and culture." By this they mean the "development of mature persons who understand the immutable connection between themselves and the land community or person/planet" and who act in ways that "serve both the vital needs of persons and nonhumans."[56]

What is common to these various responses to the ecological crisis? First and foremost, the environmental crisis is regarded not only as a crisis of participation and survival but also as a crisis of culture in the broadest sense of the term, that is, "the total of the inherited ideas, beliefs, values, and knowledge, which constitute the shared bases of social action."[57] Indeed, this was exemplified as early as 1972 in the manifesto of the New Zealand Values Party (the world's first national Green party), which spoke of New Zealand being

> in the grip of a new depression. It is a depression which arises not from a lack of affluence but almost from too much of it. It is a depression of human values, a downturn not in the national economy but in the national spirit.[58]

Second, emancipatory ecopolitical theory may be understood as challenging ecopolitical discourse and widening its agenda on three interrelated levels: human needs, technology, and self-image. At the political level, emancipatory theorists have taken the claims of the ecology movement seriously and have embarked upon a critical inquiry into the structure of human needs and the "appropriateness" of many modern technologies. It is no longer considered adequate merely to challenge, say, the site of a nuclear power plant, freeway or chemical industry, or merely to insist on better safety devices or pollution filters. Instead, this third phase of ecopolitical inquiry has sought to draw attention to the more fundamental question: to what extent do we really need these kinds of energy sources, these means of transport, these industries and technologies, and the like? Surely *more of us* (human and nonhuman) can live richer and fuller lives if humans can become less dependent on this kind of technological infrastructure and the kinds of commodities and lifestyles it offers? As Cornelius Castoriadis has observed, whereas the working class movement has mainly tackled the theme of authority (hence its focus on participatory and distributional issues), the ecology movement is now questioning

> the scheme and structure of needs and the way of life. And that constitutes a very important transcendence of what could be seen as the unilateral character of former movements. What is at stake in the ecological movement is the whole conception, the total position and relation between humanity and the world and, finally, the central and eternal question: what is human life? What are we living for?[59]

Third, this theme of cultural malaise and the need for cultural renewal has meant that emancipatory ecopolitical theorists have directed considerable attention toward the revitalization of civil society rather than, or in addition to, the state. This is reflected in the concern of emancipatory theorists to find ways of theoretically integrating the concerns of the ecology movement with other new social movements, particularly those concerning feminism, peace, and Third World aid and development. This new theoretical project is concerned to find ways of overcoming the destructive logic of capital accumulation, the acquisitive values of consumer society, and, more generally, all systems of domination (including class domination, patriarchy, imperialism, racism, totalitarianism, and the domination of nature).

This is indeed a bold and ambitious theoretical project and one for which the most influential political philosophies of modern times—conservatism, liberalism, and orthodox Marxism—appear either poorly or only partially equipped. Indeed, the limitations in these political philosophies have served as general theoretical points of departure for emancipatory ecopolitical theorists.

THE EMANCIPATORY CRITIQUE OF CONSERVATISM, LIBERALISM, AND ORTHODOX MARXISM

Although the emancipatory critique of the major political traditions has been mainly directed against liberalism and orthodox Marxism, it is useful to explore briefly the relationship between Green political thought and conservatism.[60] This is especially so since many observers on the Left have often wrongly characterized environmentalism—and, by implication, Green political thought—as simply a new incarnation of conservatism. Now it is certainly true that there are some notable points of commonality between conservatism and many strands of environmentalism. The most significant of these are an emphasis on prudence or caution in innovation (especially with respect to technology), the desire to *conserve* existing things (old buildings, nature reserves, endangered values) to maintain continuity with the past, the use of organic political metaphors, and the rejection of totalitarianism. Indeed, these links have occasionally been manifest in the appearance of ad hoc political alliances between environmental activists and traditional conservatives over specific issues such as the preservation of threatened old buildings and landscapes. Moreover, some of the political tributaries that have flowed into contemporary ecopolitical thought, and in some cases Green political thought, may be traced to conservative sources (e.g., Thomas Carlyle via William Morris, Edmund Burke via William Ophuls).[61]

Yet those who have noted the correlation between conservatism and some strands of environmentalism have acknowledged that environmentalism also contains strong elements of radicalism in its call for a rapid and far-reaching response to the current crisis. This peculiar mixture of radicalism and conservatism has understandably confounded some observers and has prompted the suggestion that perhaps environmentalists could be understood as "radical conservatives" or, more precisely, "ideational conservatives pushed into situational radicalism."[62]

However, these and other general categorizations of environmentalism cannot be simply transposed onto ecopolitical thought, least of all Green political thought. In the case of the latter, one can certainly find some of the elements of conservatism already mentioned, such as a rejection of totalitarianism, caution in technological innovation, a desire to conserve threatened buildings and landscapes, and the use of organic political metaphors. However, these elements have been rewoven into a new constellation of ideas that has a distinctly radical political edge, both in the original sense of going to the root of the problem and in the more popular sense of demanding a widespread transformation of the political and economic status quo. As we have seen, Green political thought is imbued with a culturally innovative and egalitarian ethos, which puts it at considerable odds with conservatism's opposition to social and political experimentation and cultural change and its endorsement of hierarchical authority and the established order of things. Unlike political conservatives, emancipatory ecopolitical theorists are concerned to challenge and ultimately transform existing power relations, such as those based on class, gender, race, and nationality, to ensure an equitable transition toward an ecologically sustainable society. Indeed, when it comes to modern variants of political conservatism, Greens have been some of the most vociferous critics of such neoconservative ideologies as American Reaganism, British Thatcherism, and their respective successors. Not surprisingly, emancipatory theorists have passed over conservativism (traditional and neo-) as a source of political enlightenment. Whatever the similarities to be found between conservatism and Green political thought, their fundamentally different stances with regard to power relations means that conservatism may be summarily dismissed as a serious contender in the emancipatory ecopolitical stakes.

Liberalism and Marxism, however, have attracted greater attention from emancipatory theorists, although most of this has also been critical. In particular, emancipatory theorists have done much to draw attention to the similarities between liberalism and Marxism. They have noted, for example, that while social relations between humans are theoretically different under capitalism and socialism, the relationship between humans and the rest of nature appears to be essentially the same. This has also proved to be the case historically. As Langdon Winner has remarked:

> A crucial failure in modern political thought and political practice has been an inability or unwillingness even to begin...the critical evaluation and control of our society's technical constitution. The silence of liberalism on this issue is matched by an equally obvious neglect in Marxist theory. Both persuasions have enthusiastically sought freedom in sheer material plenitude.[63]

Indeed, the international nature of environmental degradation has lent force to the broader claim by emancipatory theorists that the modern ecological crisis is the quintessential crisis of *industrialism* rather than just Western capitalism. Industrialism encompasses the "state capitalism" of communist nations as well as the largely privately controlled market capitalism of Western nations, both of which are seen by emancipatory theorists as resting upon the ideologies of growth and technological optimism. This ecological critique is therefore concerned to emphasize the shared expansionary ethos of both West and East. In the Soviet Union, this ethos was, until recently, encapsulated in the Program of the Soviet Communist Party approved in 1961 at the twenty second party Congress, which stated that "Communism elevates man to a tremendous level of supremacy over nature and makes possible a greater and fuller use of its inherent forces."[64] One could just as easily substitute Western capitalism for communism in this confident assertion of modern humanity's technological mastery of nature.

To be sure, it was classical liberalism, underpinned by *laissez faire* economics and defended in the writings of John Locke and Adam Smith, rather than communism that originally underscored the fundamental direction of modern bourgeois political economy by basing it on cornucopian assumptions and an expanding economy. As Susan Leeson has argued:

> Lockean thought legitimated virtually endless accumulation of material goods; helped equate the process of accumulation with liberty and the pursuit of happiness; helped implant the idea that with ingenuity man can go beyond the fixed laws of nature, adhering only to whatever temporary laws he establishes for himself in the process of pursuing happiness; and helped instill the notion that the "commons" is served best through each man's pursuit of private gain, because there will always be enough for those who are willing to work.[65]

Within this Lockean framework, the nonhuman world was seen in purely instrumental terms, that is, as no more than a means to human ends. After all, according to Locke, the Earth had been given to humans for "the support and comfort of their being"; moreover, the mixing of human labor with nature was an act of appropriation that created something valuable (i.e., property) out of something otherwise *valueless* (the Earth in its state of "natural grace").[66]

Of course, it must be noted that some influential liberal philosophers have challenged this instrumental and expansionary ethos and introduced important qualifications concerning the extent to which it is permissible for humans to dominate the nonhuman world. Scattered among the writings of J. S. Mill, for example, one can find a defence of ecological diversity and a brief but eloquent case for a stationary state economy.[67] And Jeremy Bentham's extension of his utilitarian calculus to all sentient beings has provided the philosophical touchstone for contemporary animal liberation theorists such as Peter Singer.[68]

Although some emancipatory theorists, such as John Rodman, have noted and discussed these byways in liberal thought, the general tendency has been to look to other political traditions for the ideals and principles that would underpin an ecologically sustainable *post-liberal* society.[69] Indeed, the classical liberal defenders of individualism and *laissez faire* economics are seen by emancipatory ecopolitical theorists as apologists for the very dynamic that has led to the "tragedy of the commons." And, as the survivalists had shown, the logical sequel of this dynamic is authoritarianism from above rather than self-limitation from below. Moreover, emancipatory theorists largely accept the democratic socialist critique of liberalism that the exercise of economic freedom by the privileged renders the exercise of both economic and political freedom largely illusory to the mass of ordinary working people, the unemployed, and the peoples of developing countries. In particular, the exercise of the inalienable rights of the individual heralded by liberalism, notably property rights (which confer the right of exclusive use and disposal of land, labor, and capital) together with freedom of contract and market incentives, is seen as leading to the concentration of ownership of capital and a system of power relations that negates the otherwise laudable liberal goal of free, autonomous development for each individual. Moreover, emancipatory theorists (like democratic socialists) do not consider it an acceptable solution merely to rely on the redistributive largesse of the welfare State to iron out excessive inequalities, since this merely brings the dispossessed into the market as passive consumers rather than self-determining producers (their only area of effective choice being how to spend their limited welfare checks). Accordingly, emancipatory ecopolitical theorists are concerned to reassert the New Left themes of participation and self-management, but in a new ecological (rather than cornucopian) context.

More importantly, liberal ideals were born in and depend upon a frontier setting and an expanding stock of wealth, with claims for distributive justice being appeased by the "trickle down" effect (which maintains relative inequalities in wealth and power). Emancipatory theorists point out that once the frontier becomes exhausted, the gap between rich and poor is bound to intensify, and the prospects of distributive justice and a more egalitarian society will become more remote.

This combined ecological and social critique of liberalism has led emancipatory theorists to reject the philosophy of possessive individualism and turn toward alternative political theories that are more consonant with an ecological perspective or, at the very least, respectful of "ecological limits," and are better able to foster some kind of democratic, cooperative, and communitarian way of organizing social and economic life.[70]

However, the orthodox Marxist alternative, while seen by many emancipatory theorists to be *theoretically* preferable to liberal political philosophy (in seeking collective economic decision making and a fairer distribution of society's stock of wealth), was found to be ultimately wedded to the same expansionary ethos and anthropocentric framework as liberalism. Moreover, as the evidence of ecological degradation in Eastern Europe mounts, communism in practice is being increasingly regarded as an unmitigated disaster from the point of view of ecological sustainability. As we shall see in chapter 4, orthodox Marxists, by and large, merely disagreed with liberals on how the drive to cornucopia was to be realized and on how the "spoils of progress" were to be managed and divided. Like Locke, Marx saw economic activity, the act of producing via the appropriation of nature, as essential to human freedom. And like Locke, Marx regarded the nonhuman world as no more than the ground of human activity, acquiring value if and when it became transformed by human labor or its extension—technology. Where Marx differed from Locke and other liberal theorists was in his rejection of the institution of private property on the grounds that it gave rise to class domination and the appropriation of surplus value from the worker.

The upshot of this critical rereading of the two most influential pillars of modern political philosophy was sobering. From Hobbes and Locke through to Marx, the notion of human self-realization through the domination and transformation of nature persisted as an unquestioned axiom of political inquiry. As Rodman has shown, in the modern era the solution to poverty, injustice, and inequality had become dependent on the abolition of scarcity via technological innovation and industrial growth—an approach that has been traced to the Enlightenment ideal of the progressive liberation of humans from all traditional and natural limits.[71] Now, however, emancipatory theorists have carried forward the survivalist argument that the modern era must be seen as but a temporary suspension of the tradition of scarcity, as an aberrant period in human history. Some have likened it to the "pioneer" stage of ecological succession (i.e., where rapid growth and aggressive exploitation takes place), which must soon phase into a more mature, steady-state, climax community.[72]

Although emancipatory theorists are in general agreement that liberalism and orthodox Marxism provide unsuitable theoretical underpinnings for an ecologically benign, conserver society, they differ markedly on the question of alternatives. As we shall see in chapters 4 to 7, this new breed of ecologically oriented theorists rapidly divided over the question as to what kind of post-liberal social and political theory could best address the interrelated social and environmental problems of the modern world: was it neo- or post-Marxism, democratic socialism, utopian socialism, anarchism, feminism, or some revised combination thereof?

At a more fundamental ecophilosophical level, deep divisions also developed over the question of our proper relationship to the nonhuman world. While most emancipatory theorists agree that it is not enough simply to return to the participatory and countercultural ethos of the 1960s (with its cornucopian assumptions of an ever growing stock of wealth), serious disagreement developed as to *how far* the anthropocentric assumptions and technological

153,342

aspirations of the modern world needed to be revised. This has given rise to *the* most fundamental division within emancipatory ecopolitical thought.

THE ANTHROPOCENTRIC/ECOCENTRIC CLEAVAGE
WITHIN EMANCIPATORY THOUGHT

It should be clear from the above brief introduction to emancipatory inquiry that it is best understood as representing a *spectrum* of thought rather than a single ecopolitical theory or an internally coherent bundle of ideas—a situation that reflects the current state of day-to-day Green politics. Although there are many different areas of disagreement, the most fundamental division from an ecophilosophical point of view is between those who adopt an anthropocentric ecological perspective and those who adopt a nonanthropocentric ecological (or ecocentric) perspective. The first approach is characterized by its concern to articulate an ecopolitical theory that offers new opportunities for *human* emancipation and fulfilment in an ecologically sustainable society. The second approach pursues these same goals in the context of a broader notion of emancipation that also recognizes the moral standing of the nonhuman world and seeks to ensure that it, too, may unfold in its many diverse ways. This anthropocentric/ecocentric cleavage follows the ecophilosophical cleavage that is central to the relatively new but rapidly expanding field of environmental philosophy. The centrality of this distinction is reflected in the large number of broadly similar distinctions that have been coined not only in ecopolitical thought and environmental philosophy but also in environmental history and environmental sociology. It is reflected, for example, in Arne Naess's influential distinction between shallow ecology and deep ecology; in Timothy O'Riordan's characterization of "technocentrism" and "ecocentrism"; in the "Imperialist" and "Arcadian" traditions of ecological thought identified by the environmental historian Donald Worster; in Murray Bookchin's distinction between "environmentalism" and "social ecology"; in William Catton and Riley Dunlap's distinction between the dominant "Human Exemptionalism Paradigm" of mainstream sociology and the "New Ecological Paradigm" of the "post-exuberant age"; and in Alan Drengson's distinction between the "technocratic" and "pernetarian" (i.e., person-planetary) paradigms.[73]

Although some of these distinctions bear different nuances, they all contrast a human-centered orientation toward the nonhuman world with an ecology-centered orientation. In the case of the former, the nonhuman world is reduced to a storehouse of resources and is considered to have instrumental value only, that is, it is valuable only insofar as it can serve as an instrument, or as a means, to human ends. The latter approach, on the other hand, also values the nonhuman world—or at least aspects of it—for its own sake.

While Naess's brief but fertile characterization of deep and shallow ecology has proved to be the most influential in ecophilosophical circles, I will use the more general ecocentric/anthropocentric distinction for the purposes of this inquiry since it is more immediately descriptive of the two opposing orientations it represents.[74] Deep ecology, or (after Fox) "transpersonal ecology," may be understood as representing one very promising and distinctive *kind* of ecocentric approach (transpersonal ecology and other examples of ecocentric approaches are discussed in chapter 3).[75]

An alternative approach to classification might have been to locate emancipatory theory on the familiar left/right political spectrum. However, as we have seen, most contributors to this third phase of ecopolitical inquiry tend, in any event, to cluster to the left of this traditional spectrum insofar as they are seeking some kind of communitarian, cooperative, or democratic socialist solution (and here, it is not clear which of these approaches are supposed to be "more to the left"). Its use as an analytical framework in this context is therefore decidedly limited.

Another dimension that might be more profitably applied to these various left-leaning emancipatory approaches is that of community versus state control. In terms of our tripartite characterization of ecopolitical theory, this dimension would shed light on the different attempts by emancipatory theorists to resolve the tension between the participatory and survivalist themes of ecopolitical thought already discussed. It would also bring into sharp relief the differences between emancipatory theorists on matters such as political organization and strategy. However, as important as these themes are to Green political theory (particularly with respect to the debates between ecoanarchists and ecosocialists, as we shall see in chapters 6 and 7), the community versus state control dimension does not highlight what is distinctive about the emancipatory approach (i.e., the emphasis on cultural renewal, the emphasis on developing an ecological consciousness, and the critique of industrialism). More importantly, such a dimension does not adequately register the major *ecophilosophical* debates in emancipatory thought. Nonetheless, the community versus state control dimension can serve as a useful adjunct to the more overarching ecophilosophical dimension.

The anthropocentric/ecocentric dimension registers the major ecophilosophical differences within emancipatory ecopolitics and brings into sharp focus the novel and challenging scope of these new ideas. Moreover, it does this in a way that helps to *explain* some of the diverging *political* responses to different ecological issues adopted by different schools of emancipatory thought, as I show below in my discussion of what I identify as two "litmus test" ecological issues.

For the reasons developed in the next two chapters, I will be arguing that an ecocentric philosophical orientation provides the most comprehensive, promising, and distinctive approach in emancipatory ecopolitical theory. Accordingly, the various Green political theories examined in part 2 of this inquiry will be assessed in terms of where they fit on the anthropocentric/eco-centric emancipatory dimension. To the extent that they fall short of a com-prehensive ecocentric perspective, they will be judged inadequate. To the extent that they point to problems associated with an ecocentric perspective, their critique will be addressed and evaluated. And to the extent that they con-tribute to the rounding out or further elaboration of an ecocentric political per-spective, particularly on social and institutional questions where much work needs to be done, their contribution will be incorporated accordingly.

What, then, are the salient features of the ecocentric approach? In terms of fundamental priorities, an ecocentric approach regards the question of our proper place in the rest of nature as logically prior to the question of what are the most appropriate social and political arrangements for human communi-ties. That is, the determination of social and political questions must proceed from, or at least be consistent with, an adequate determination of this more fundamental question. As exemplified in some of the quotations selected to introduce this third phase of inquiry, ecocentric political theorists are distin-guished by the emphasis they place on the need for a radical reconception of humanity's place in nature. In particular, ecocentric theorists argue that there is no valid basis to the belief that humans are the pinnacle of evolution and the sole locus of value and meaning in the world. Instead, ecocentric theo-rists adopt an ethical position that regards *all* of the various multilayered parts of the biotic community as valuable for their own sake. (There are, of course, different degrees of anthropocentrism and ecocentrism, as will be seen in the following chapters. Here I am simply characterizing a thorough-going ecocentric perspective.)

The special emphasis given to ecological interconnectedness by ecocen-tric theorists is seen as providing the basis for a new sense of both *empathy* and *caution*. By this I mean a greater sense of compassion for the fate of other life-forms (both human and nonhuman) and a keener appreciation of the fact that many of our activities are likely to have a range of unforeseen consequences for ourselves and other life-forms.[76] The magnitude of the environmental crisis is seen by ecocentric theorists as evidence of, among other things, an inflated sense of human self-importance and a misconceived belief in our capacity to fully understand biospherical processes. The ecocen-tric perspective is presented as a corrective to these misconceptions insofar as it underscores the need to proceed with greater caution and humility in our "interventions" in ecosystems.

It was the adoption of this thoroughgoing ecocentric perspective that most set this particular group of emancipatory ecopolitical thinkers apart from most of the influential New Left theorists of the 1960s who had addressed the problem of environmental degradation. To be sure, there has been an important re-assertion by ecocentric theorists of New Left themes (such as autonomy, self-management, and the critique of technocratic rationality) in response to authoritarian ecopolitical solutions. However, these themes have been relocated in a new ecocentric theoretical framework that draws inspiration from the insights of ecology rather than from the human-centered orientation of the New Left. Anthropocentric Green theorists, on the other hand, have maintained greater continuity with the New Left themes of the 1960s. The main point of difference, however, is that anthropocentric Green theorists have revised the cornucopian assumptions of the 1960s in the wake of the "limits to growth debate" of the early 1970s. The result is a more ecologically informed (albiet still human-centered) emancipatory theory that provides a much more comprehensive critique of economic growth and technocratic rationality.

Ecocentric and anthropocentric emancipatory theorists offer diverging responses to a range of important practical social and ecological issues. In particular, I would point to two "litmus" ecopolitical issues that highlight these ecophilosophical differences: human population growth and wilderness preservation. The ecocentric stream of emancipatory thought is noted for its greater willingness to advocate not simply a lessening of the growth rate of the human population but also a long term *reduction* in human numbers.[77] Rather than directly address the matter of absolute numbers, the anthropocentric stream tends to direct attention to the social causes of population growth and argue the case for a more equitable distribution of resources between the rich and poor. The ecocentric stream is also noted for its greater readiness to advocate the setting aside of large tracts of wilderness, regardless of whether such preservation can be shown to be useful in some way to humankind. The anthropocentric stream, in contrast, tends to be more preoccupied with the urban and agricultural human environment. Large scale wilderness preservation tends not to be supported unless a strong human-centered justification can be demonstrated.

The ecophilosophical differences between ecocentric and anthropocentric theorists should not, of course, obscure the significant commonalities between these two streams of emancipatory thought. As we have seen, both streams are distinguishable from other ecopolitical approaches in terms of their more penetrating diagnosis of environmental problems (i.e., these are seen as representing not just a crisis of participation and survival but also a crisis of culture and character). Both streams are also united in their optimistic attempt to offer a creative synthesis of the themes of participation and survival through the more encompassing theme of emancipation, which promises new opportunities for *universal* human self-realization. At the policy level, both streams are critical of indiscriminate economic growth, large scale organizations, "hard" (as distinct from "soft") energy paths, and ecologically and socially destructive technologies. Where these two approaches differ, however, is in the way in which they integrate these critiques and in the ecophilosophical justifications they provide for their alternative approaches.

Having located the ecocentric emancipatory stream in the larger body of ecopolitical thought, the central questions to be examined in this inquiry can now be presented: (i) does an ecocentric approach have a natural ally within the existing pantheon of modern political traditions with which it can forge a theoretical link; or (ii) can an ecocentric approach be assimilated into any one of a number of different political traditions after appropriate revisions; or (iii) must ecocentric theorists develop an entirely novel political arrangement?

In order to narrow down the field of choice, it will be useful at this stage to outline a response to these questions from the perspective of emancipatory ecopolitical thought in general. This will provide the general parameters for the ensuing inquiry.

Although there is at present no unanimity among emancipatory theorists in response to these questions, definite leanings are discernible. First, as we have seen, emancipatory theorists are united by their intention to "head off" the acknowledged possibility of the survivalist solution, namely, that only a centrally planned, authoritarian State is capable of steering modern industrialized society through the convulsive process of de-industrialization into an ecologically sustainable, post-industrial society.

Second, the conservative political tradition may be ruled out as a serious contender, notwithstanding the resonances with emancipatory ecopolitical thought that have been briefly noted in this chapter. This is because conservatism's endorsement of the established order, hierarchical authority, and paternalism and its resistance to cultural innovation and social and political experimentation put it at considerable odds with the egalitarian and innovative orientation of emancipatory thought.

Third, all emancipatory theorists roundly reject "free market" liberalism and neoconservatism as giving free rein to the very dynamic that has given rise to the "tragedy of the (unmanaged) commons." This does not entail an outright rejection of entrepreneurial activity or of the market as a method of resource allocation, but it does require that the market become subordinate to ecological and social justice considerations. Beyond this, however, emancipatory theory, particularly the ecocentric stream, is still very much in its infancy and there is, so far, little agreement as to what mix of private and public economic endeavor would best secure a socially just and ecologically sustainable society. The arguments for the rejection of classical liberal philosophy have already been canvassed earlier in this chapter and will not be pursued in any detail in the remainder of this inquiry. It should be noted, however, that the emancipatory critique of liberalism has not led to an outright rejection of the entire cluster of liberal values. The (usually unacknowledged) retention by emancipatory theorists of the enduring liberal values of tolerance for diversity, basic human rights (e.g., freedom of speech, assembly, and association), and (for some) limited government indicates that emancipatory political theory is decidedly *post-* rather than *anti*-liberal.

Fourth, although Marxist and neo-Marxist theories have also attracted their due share of ecological critiques, they have, on the whole, proved to be more resilient than classical liberal approaches. Marxism's penetrating critique of capitalist relations and its promise of universal human self-realization has continued to exert a considerable sway on the anthropocentric and, to a much lesser extent, ecocentric streams of emancipatory thought. For these reasons, Marxist and neo-Marxist responses will be critically explored in detail in chapters 4 and 5 (if only to demonstrate why both are ultimately incompatible with an ecocentric perspective).

Fifth, in view of the broad egalitarian and democratic ethos of emancipatory thought and its sympathy with the concerns of new social movements, feminist, democratic socialist, utopian socialist, and anarchist approaches have enjoyed widespread support among emancipatory theorists of both persuasions. Accordingly, these political theories (or ecophilosophies, in the case of ecofeminism) and will be examined in chapters 3, 6, and 7.

Sixth, no emancipatory theorist has been able to come up with an entirely novel social and political arrangement, that is, one that has not already been mooted in modern social and political theory. By this I am not meaning to argue that there is nothing new or distinctive about Green political thought, only that the newness or distinctiveness of Green political thought is not primarily to be found in the various social and political insititutions defended by its theorists.[78] Rather, the principal newness or distinctiveness of emancipatory thought (and this applies more to the ecocentric than the anthropocentric stream) lies in the different ecophilosophical perspective that is brought to bear upon contemporary problems, the different and more encompassing kind of critique that is applied to existing social and political institutions, and the different and more encompassing ethical and political justifications provided for the various (not unfamiliar) social and political arrangements that are proposed.

CHAPTER 2

Exploring the Environmental Spectrum: From Anthropocentrism to Ecocentrism

INTRODUCTION

Although an ecological perspective is basic to Green political thought, there is a diversity of views among Green theorists as to the meaning, scope, and political consequences of such a perspective. As we saw in chapter 1, the most fundamental area of divergence is between an anthropocentric and an ecocentric orientation. However, these two ecological orientations merely represent the opposing poles of a wide spectrum of differing orientations toward nature. This chapter will be drawing on recent work in environmental philosophy to provide an overview and discussion of the major streams of modern environmentalism, most of which fall *between* these two poles. This discussion will provide the conceptual tools that will enable an evaluation of the particular *kind* and *degree* of anthropocentrism or ecocentrism that is manifest or latent in the various Green political theories to be examined in part 2.

MAJOR STREAMS OF ENVIRONMENTALISM[1]

In presenting the following overview of the major streams of environmentalism, I have drawn on the pioneering typologies of environmentalism developed by John Rodman and, more recently, Warwick Fox, who elaborates the most exhaustive classification scheme in the ecophilosophical literature.[2] The concern of these thinkers has been to characterize the major arguments and problems associated with different environmental positions in order to distinguish and defend an "ecological sensibility" as the basis for a general environmental ethic (Rodman) or a similar, but more detailed, "transpersonal ecology" approach to ecophilosophy (Fox). Whereas Rodman has sought to crystallize the major currents in the history of the environmental movement in order to uncover their complexities and ambiguities, Fox has developed a more general, analytical map that is intended to provide a close to exhaustive categorization of the range of ecophilosophical positions (i.e., whether or not they are represented by a particular historical movement).[3] The approach here will be primarily historical, since my main concern is to relate clusters of particular environmental ideas to particular movements and to point out the contribution of, ambiguities in, and potential for alliance between, these various movements. Above all, I am concerned to identify the major currents of contemporary environmentalism that have seeped, in varying degrees, into the central "ecological pillar" of Green politics. It is very important to undertake such a survey in a political inquiry of this kind for the simple reason that most Green political theorists (as distinct from ecophilosophers) have so far paid insufficient attention to articulating the ambit of the central pillar of ecology in any kind of detail or to exploring the social and political implications of different kinds of environmental postures.

Moving from the anthropocentric toward the ecocentric poles, the major positions that I will be discussing are resource conservation, human welfare ecology, preservationism, animal liberation, and ecocentrism.[4] This spectrum represents a general movement from an economistic and instrumental environmental ethic toward a comprehensive and holistic environmental ethic that is able to accommodate human survival and welfare needs (for, say, a sustainable "natural resource base," a safe environment, or "urban amenity") while at the same time respecting the integrity of other life-forms. However, since part of my concern is to draw out the ambiguities in, and the potential for forming alliances between, some of these historical currents of environmentalism, the general movement from anthropocentrism to ecocentrism will not appear as a strict linear progression. For example, some of the arguments for preservationism are *more* ecocentric than those for animal liberation, while other preservationist arguments represent a variation of some of the arguments used by the human welfare ecology stream.

This general overview of environmentalism will also help to explain how some currents of environmentalism have had more influence in some countries than others and how this has influenced both the nature and goals of the Green movement and the expression of Green theory in those countries. For example, the human welfare ecology stream has played a relatively more prominent role in Europe, whereas the preservationist stream has had more influence in "New World" regions such as North America, Australia, and New Zealand (where there are considerably more areas of wilderness to preserve).[5] This has given rise to different emphases in Green theory and practice in those regions. Moreover, in part 2 we shall see how those environmental streams clustering toward the anthropocentric end of the spectrum can be more readily accommodated within existing political traditions, whereas those clustering toward the ecocentric end cannot easily mesh with such traditions, at least not in the absence of major theoretical revisions.

There is, of course, a considerable overlap between the various currents of environmentalism to be discussed in terms of their practical upshot. However, we shall see that these currents vary markedly in their comprehensiveness and philosophical basis and that this has important implications when it comes to deciding which perspective is best able to provide the ecophilosophical touchstone for Green political thought.

Resource Conservation

Although the idea of conservation, in the sense of the "prudent husbanding" of nature's bounty, can be traced back as far as Plato, Mencius, Cicero, and the Old and New Testaments, its twentieth-century scientific and utilitarian manifestation is intricately bound up with the rise of modern science from the sixteenth century.[6] Those who have inquired into the historical roots of the modern conservation doctrine have generally traced its popularization in North America to Gifford Pinchot, the first chief of the United States Forest Service, described by Devall as the "prototype figure in the [conservation] movement."[7] Central to Pinchot's notion of conservation was the elimination of waste, an idea that the environmental historian Samuel P. Hays has dubbed "the gospel of efficiency," which he sees as lying at the heart of the doctrine of conservation. Yet Pinchot's ideas were also deeply imbued with the ethos of the Progressive era to which he belonged; indeed, in his book *The Fight for Conservation,* he identified "development" as the first principle of conservation, with "the prevention of waste" and development "for the benefit of the many, and not merely the profit of the few" forming the second and third principles, respectively.[8] Moreover, as McConnell observes, it was taken for granted that the principle of waste prevention meant "maximizing output of economic goods per unit of human labor."[9] According to Devall, the Pinchot-led conservation movement in the United States helped to "professionalize 'resource management'" and further the centralization of power in large public bodies (such as the U.S. Forest Service) based on principles of "scientific management."[10]

Rodman has labelled this modern scientific and utilitarian approach to land management the "resource conservation" movement and has described it as "an unconstrained total-use approach, whose upshot is to leave nothing in its natural condition (for that would be a kind of 'waste,' and waste should be eliminated)."[11] Similarly, Devall and Fox refer to this perspective as the "resource conservation *and development*" (my emphasis) perspective in order to underscore the point that waste meant not only the inefficient use of natural resources but also their *nonuse.*

The resource conservation perspective may be seen as the first major stop, as it were, as one moves away from an unrestrained development approach. Not surprisingly, it is the least controversial stream of modern environmentalism—indeed, it has become somewhat of a foe to more radical streams of environmentalism. The general acceptability of the resource conservation perspective arises from the fact that it proceeds from a human-centered, utilitarian framework that seeks the "greatest good for the greatest number" (including future generations) by reducing waste and inefficiency in the exploitation and consumption of nonrenewable "natural resources" (e.g., oil) and ensuring a maximum sustainable yield in respect of renewable resources (e.g., fisheries, soil, crops, and timber). As such, it is a perspective that is inextricably tied to the production process and, by virtue of that fact, necessarily regards the nonhuman world in use-value terms. This is reflected, among other things, in the *language* used by adherents of this stream of environmentalism; after all, "resources" are, as Neil Evernden points out, "indices of utility to industrial society. They say nothing at all of experiential value or intrinsic worth."[12] Similarly, Laurence Tribe has argued that to treat human material satisfaction as the *only* legitimate referent of environmental policy analysis and resource management leads to "the dwarfing of soft variables" such as the aesthetic, recreational, psychological, and spiritual needs of humans and the different needs of *other* life-forms.[13] While the recognition of the use value of the nonhuman world must form a necessary part of any comprehensive environmental ethic, resource conservation is too limited a perspective to form the *exclusive* criterion of even a purely anthropocentric environmental ethic.

Human Welfare Ecology

Like the resource conservation stream, the movement for a safe, clean, and pleasant human environment has a long pedigree, although the pace, reach, and expectations of such concern has grown considerably since the onset of the industrial revolution, and even more so since the 1960s. Whereas the labor movement had been in the forefront of the early wave of demands for a safer and more agreeable work environment (and Engels's classic 1845 critique of the conditions of the Victorian working class must be seen as a major milestone in the development of this movement), the late twentieth century bearers of this human welfare stream have increasingly been citizens, consumers, and "householders" concerned with the state of the residential and urban environment.[14] This is reflected in the increasing role played by women in urban ecological protest and in the changing sites of political struggle—from the factory to the household, street, shopping mall, and local municipal government. That human welfare ecology protest may appear today to be a peculiarly late twentieth century phenomenon may be attributed as much to the rapid escalation in urban and agricultural environmental problems since World War II as to the emergence of "post-material" values borne by the so-called New Middle Class.[15] The accumulation of toxic chemicals or "intractable wastes"; the intensification of ground, air, and water pollution generally; the growth in new "diseases of affluence" (e.g., heart disease, cancer); the growth in urban and coastal high rise development; the dangers of nuclear plants and nuclear wastes; the growth in the nuclear arsenal; and the problem of global warming and the thinning of the ozone layer have posed increasing threats to human survival, safety, and well-being.

The goals of the human welfare ecology stream for a cleaner, safer, and more pleasing human environment are relatively straightforward and represent a more generalized form of prudence and enlightened self-interest than the resource conservation stream. Indeed, they provide an important challenge to the narrow, economistic focus of resource conservationists. Whereas the resource conservation movement has been primarily concerned with improving economic productivity by achieving the maximum sustainable yield of natural resources, the major preoccupation of the human welfare ecology movement has been the health, safety, and general amenity of the urban and agricultural environments—a concern that is encapsulated in the term "environmental quality."[16] In other words, the resource conservation stream may be seen as primarily concerned with the *waste* and *depletion* of natural resources (factors of production) whereas the human welfare ecology stream may be seen as primarily concerned with the *general degradation,* or overall state of health and resilience, of the physical and social environment. For the human welfare ecology stream, then, "sustainable development" means not merely sustaining the natural resource base for human *production* but also sustaining biological support systems for human *reproduction.* Moreover, by focusing on both the physical *and* social limits to growth, the human welfare ecology stream has done much to draw attention to those "soft variables" neglected by the resource conservation perspective, such as the health, amenity, recreational, and psychological needs of human communities.

More significantly, the human welfare ecology stream, unlike the resource conservation stream, has been highly critical of economic growth and the idea that science and technology alone can deliver us from the ecological crisis (although the human welfare ecology stream has, of course, been dependent on the findings of ecological science to mount its case). The kind of ecological perspective that has informed this stream of environmentalism is encapsulated in Barry Commoner's "four laws of ecology": everything is connected to everything else, everything must go somewhere, nature knows best (i.e., any major human intervention in a natural system is likely to be detrimental to that system), and there is no such thing as a free lunch.[17] These popularly expressed ecological insights have challenged the technological optimism of modern society and the confident belief that, in time, we can successfully manage all our large-scale interventions in natural systems without any negative consequences for ourselves. The realization that there is no "away" where we can dump our garbage, toxic and nuclear wastes, and other kinds of pollution has given rise to calls for a new stewardship ethic— that we must protect and nurture the biological support system upon which we are dependent. Practically, this has led to widespread calls for "appropriate technology" and "soft" energy paths, organic agriculture, alternative medicine, public transport, recycling, and, more generally, a revaluation of human needs and a search for more ecologically benign lifestyles.

Since it is in urban areas that we find the greatest concentration of population, pollution, industrial and occupational hazards, traffic, dangerous technologies, planning and development conflicts, and hazardous wastes, it is hardly surprising that cities and their hinterlands have provided the major locale and focus of political agitation for human welfare ecology activists. Nor is it surprising that human welfare ecology has been the strongest current of environmentalism in Green politics in the most heavily industrialized and domesticated regions of the West, most notably Europe. In particular, the many different popular environmental protests or "citizen's initiatives" in West Germany that provided the major impetus to the formation of *Die Grünen* have primarily been urban ecological protests falling within this general rubric. Not surprisingly, the ecological pillar in *Die Grünen*'s platform is generally couched in the language of human welfare ecology.[18]

By virtue of its primary concern for *human welfare* in the domestic environment, however, this stream has generally mounted its case on the basis of an anthropocentric perspective. That is, the public justification given for environmental reforms by human welfare ecology activists has tended to appeal to the enlightened self-interest of the human community (e.g., for *our* survival, for *our* children, for *our* future generations, for *our* health and amenity). Indeed, the human welfare ecology stream has no need to go any further than this in order to make its case: it is enough to point out that "we must look after nature because it looks after *us*." Moreover, defenders of this perspective can say to their ecocentric critics that human welfare ecology reforms would, in any event, directly improve the well-being of the nonhuman community as well. Why, they ask, should we challenge the public and lose the support of politicians with perplexing and offbeat ideas like "nature for its own sake" when we can achieve substantially the same ends as those sought by ecocentric theorists on the basis of our own mainstream anthropocentric arguments? The ecocentric rejoinder, however, is that if we restrict our perspective to a human welfare ecology perspective we can provide no protection to those species that are of no present or potential use or interest to humankind. At best wildlife might emerge as an indirect beneficiary of human welfare ecology reforms.[19] More generally, an anthropocentric framework is also likely to wind up reinforcing attitudes that are detrimental to the achievement of comprehensive environmental reform in the long run, since human interests will always systematically prevail over the interests of the nonhuman world. As Fox puts it, employing only anthropocentric arguments for the sake of expediency might win the occasional environmental battle in the short term. However, in the long term "one is contributing to losing the ecological war by reinforcing the cultural perception that what is valuable in the nonhuman world is what is useful to humans."[20]

Preservationism

If the essence of the resource conservation stream is the "wise-use" of natural resources, and the essence of the human welfare ecology stream is the pursuit of environmental quality, then the essence of the early preservationist stream may be described as reverence, in the sense of the aesthetic and spiritual appreciation of wilderness (i.e., nonhuman nature that has not, or only marginally, been domesticated by humans). In North American environmental history, the conflict between Gifford Pinchot of the U.S. Forest Service, on the one hand, and John Muir of the Sierra Club, on the other hand, is generally taken as the archetypical example of the differences between resource conservation and preservation. In short, whereas Pinchot was concerned to *conserve* nature *for* development, Muir's concern was to *preserve* nature *from* development.[21]

The precedent for the reservation of large wilderness areas was set in the latter half of the nineteenth century, the most significant milestone being the designation of over two million acres of northwestern Wyoming as Yellowstone National Park in 1872. According to Nash, this designation was "the world's first instance of large-scale wilderness preservation in the public interest."[22] However, similar developments were also occurring in Australia; the eighteen thousand acre Royal National Park, near Sydney (set aside in 1879 "for the use of the public forever as a national park"), is often cited as the second oldest national park.[23] Although the early reservations were made primarily in order to preserve "scenery" and provide recreational facilities for public use, the twentieth century has witnessed a considerable broadening of the case for preservation along with its base of popular support.

It is noteworthy that, whereas wilderness was once feared by the early European colonists in New World regions such as Australia and North America as a hostile force to be tamed, to an increasing number of Westerners wilderness has become, for a complex range of reasons, a subject of reverence, enlightenment, and a locus of threatened values. The recent success of the Tasmanian Wilderness Society's campaign to "save" the Franklin river from a proposed dam by the Hydro-Electric Commission of Tasmania is one of the latest in a series of preservationist campaigns that have drawn support from a growing wellspring of popular sentiment and concern for the flourishing of pristine wilderness.[24] Indeed, it is arguably the campaigns for wilderness preservation, more than any other environmental campaigns, that have generated the most radical philosophical challenges to stock assumptions concerning our place in the scheme of things, thereby forcing theorists to confront the question of the moral standing of the nonhuman world. Despite John Muir's pious and outmoded vocabulary, his public defence of "wild nature" has made a lasting impression on the modern environmental imagination. As Stephen Fox shows, Muir found a divergence between Christian cosmology and the evidence of nature. "The world we are told was made for man," he noted. "A presumption that is totally unsupported by facts.... Nature's object in making animals and plants might possibly be first of all the happiness of each one of them, not the creation of all for the happiness of one. Why ought man to value himself as more than an infinitely small composing unit of the one great unit of creation?... The universe would be incomplete without man; but it would also be incomplete without the smallest transmicrosopic creature that dwells beyond our conceitful eyes and knowledge."[25]

The link between Muir's particular pantheistic worldview and the ecocentric philosophy of more recent times is widely acknowledged, although there are important differences. Rodman, for example, has argued that Muir's egalitarian orientation toward other species was "faint in comparison to the religious/esthetic theme" in his life and writings—and that an ethic that is primarily based on awe has significant limitations.[26] Insofar as wilderness appreciation has developed into a cult in search of sublime settings for "peak experiences" or simply places of rest, recreation, and aesthetic delight—"tonics" for jaded Western souls—it tends to converge with the resource conservation and human welfare ecology positions in offering yet another kind of human-centered justification for restraining development. Moreover, this kind of preservationism has sometimes been unduly selective in that it has traditionally tended to single out those places that are aesthetically appealing according to Western cultural mores (e.g., pristine lofty mountains, grand canyons, and wild rivers). These areas are often considered holier and therefore more worthy of being "saved" than places that lack the requisite grandeur or sublime beauty (e.g., wetlands, degraded farm land, roadside vegetation)—even though the latter may be more ecologically significant or contain threatened species. This trend has been gradually reversed, however, as the preservationist movement has become more ecologically informed and flowered into a thoroughgoing ecocentric environmentalism.

Moreover, J. Baird Callicott—a thoroughgoing nonanthropocentric philosopher—has criticized the very concept of wilderness on the grounds that it enshrines a bifurcation between humanity and nature; is ethnocentric and sometimes racist (e.g., it overlooks the ecological management by the aboriginal inhabitants of New World "wilderness" regions through such practices as fire lighting) and ignores the dimension of time in suggesting that the ecological status quo should be "freeze-framed."[27] As Callicott rightly argues, we need to be wary of reinforcing this human/nature bifurcation and to develop, where possible, a more dynamic and symbiotic approach to land management that acknowledges that humans can live alongside wild nature. However, as Callicott acknowledges, there also remains a strong case for the reservation and protection of large areas of representative ecosystems as the best means of conserving species diversity and enabling ongoing speciation.[28]

Finally, from an ecological point of view, it is self-defeating to focus exclusively on setting aside pockets of pristine wilderness while ignoring the growing problems of overpopulation and pollution, since these problems will sooner or later impact upon the remaining fragments of wild nature. As Rodman has observed, "the logic of preserving wilderness and wildlife on artificial islands surrounded by the sea of civilization seems to involve its own mode of destruction."[29] In this respect, the human welfare ecology movement provides an essential complement to the preservationist movement, as most contemporary wilderness activists recognize. As we have already noted, most of the environmental reforms pursued by human welfare ecology activists help, albeit indirectly, to secure the ecological integrity of wilderness areas (e.g., by minimizing resource use and pollution). Indeed, the overarching problem of global warming has now prompted a considerable rallying together of different environmental groups in recognition of their mutual interest in minimizing climatic disruption.

More recently, environmental philosophers have pointed to the wide range of anthropocentric arguments that have been advanced in favor of wilderness preservation (some of which have already been canvassed above). Fox provides the most exhaustive classification of these arguments to date.[30] Building on and adding to work by William Godfrey-Smith and George Sessions, Fox identifies nine kinds of argument for preserving the nonhuman world on the basis of its instrumental value to humans. He refers to these as the "life-support," "early warning system," "laboratory" (i.e., scientific study), "silo" (i.e., stockpile of genetic diversity), "gymnasium" (i.e., recreational), "art gallery" (i.e., aesthetic), "cathedral" (i.e., spiritual), "monument" (i.e., symbolic) and "psychogenetic" (i.e., psychological health and maturity) arguments. He also divides these nine arguments into five general categories of argument that emphasize the "physical nourishment value," the "informational value," the "experiential value," the "symbolic instructional value," and the "psychological nourishment value" of the nonhuman world to humans.

It is easy to see how many of the more tangible arguments for the preservation of wilderness can be quite persuasive politically, especially the more economically inclined arguments, such as those that refer to the recreational potential of wilderness or those that demonstrate the importance of maintaining genetic diversity to provide future applications in medicine and agriculture. However, it is important not to underestimate the political potency of some of the less tangible arguments for wilderness preservation. For example, the preservation of wild nature is seen by many as both a symbolic act of resistance against urban and cultural monoculture and the materialism and greed of consumer society *and* a defence (both real and symbolic) of a certain cluster of values. These include freedom, spontaneity, community, diversity, and, in some cases, national identity.[31]

Part of the political potency of arguments of this latter kind lies in the fact that the defence of wild nature is at the same time a defence of a certain cluster of values of *social* consequence. That is, they represent not only a defence of biological diversity and of "letting things be" but also a renewed assault on the one-dimensionality of technological society. In this respect, Thoreau's oft-quoted dictum—"in wilderness is the preservation of the world"—may be seen as taking on both an ecological *and* political meaning.

While many of the arguments discussed above are essentially instrumental and anthropocentric (since they are primarily concerned with defending the material and experiential benefits of wilderness to humankind), some also address deep-seated questions concerning human identity in a way that has invited a shift in our general orientation toward the world, both human and nonhuman. This is because examining our relationship to other life-forms tells us something about ourselves—about our modern character and the kinds of values and dispositions that our society encourages or discourages.[32] As we have seen, the most radical argument to emerge from this kind of ecophilosophical soul searching—an argument foreshadowed by Muir—is that we should value nature not only for its instrumental value to *us* but also for its *own* sake. It is in this particular respect that the preservationist stream of environmentalism may be seen as the harbinger of ecocentrism.

Animal Liberation

Alongside the three major streams of environmentalism discussed above is a fourth stream that has developed relatively independently and has its origins in the various "humane" societies for the prevention of cruelty to animals that emerged in the eighteenth and nineteenth centuries. The modern animal liberation movement, unlike the resource conservation, human welfare ecology, and preservation movements, has from its inception consistently championed the moral worthiness of certain members of the nonhuman world.[33] However, while the animal liberation movement might have been one of the first streams of environmentalism to have stepped unambiguously over what might be called the "great anthropocentric divide," such a step, as many ecophilosophical critics have recently pointed out, was not as momentous as it might first appear. In the view of these critics, the philosophical foundations of the animal liberation movement are unduly limited and fall well short of a rounded ecocentric worldview.[34]

The popular case for the protection of the rights of animals is a relatively straightforward revival of the arguments of the modern utilitarian school of moral philosophy founded by Jeremy Bentham. In enlarging the conventional domain of ethical theory, Bentham had argued that human moral obligation ought to extend to all beings capable of experiencing pleasure and pain, regardless of what other characteristics they may possess or lack. The important question for Bentham in respect of whether beings were morally considerable was "not, Can they *reason*? nor, Can they *talk*? but, *Can they Suffer*?"[35]

In drawing on Bentham's moral philosophy, the contemporary animal rights theorist Peter Singer has argued in favor of the moral principle of equal consideration (as distinct from treatment) of the interests of all sentient beings regardless of what kind of species they are.[36] The criterion of sentience is pivotal. For example, Singer has insisted that the "capacity for suffering and enjoyment is *a prerequisite for having interests at all,* a condition that must be satisfied before we can speak of interests in a [morally] meaningful way"—indeed, he has argued that the criterion of sentience it is the "only defensible boundary of concern for the interests of others."[37] (The question as to what is a sentient being is not always easy to determine. Singer has sought to show that there is ample evidence that mammals, birds, reptiles, fish and, to a lesser extent, crustaceans all feel pain. He concedes that determining the exact cutoff point is difficult but suggests that "somewhere between a shrimp and an oyster seems as good a place to draw the line as any, and better than most.")[38]

To Singer, it is morally irrelevant whether a being possesses such capacities as linguistic skills, self-consciousness, or the ability to enter into reciprocal agreements (which represent some of the usual kinds of justification given for according humans exclusive moral standing), if that being is otherwise sentient. In this respect, Singer has done much to expose the logical inconsistency in the practice of taking into account and protecting the interests of handicapped or immature humans such as brain damaged people, the senile, or infants, yet continuing to ignore the suffering imposed on nonhuman animals in such practices as "factory farming" and vivisection. After all, as Singer provocatively asks, if there are some handicapped humans who are no more rational than nonhuman animals, then why not use *them* in scientific experimentation? By analogy with racism, Singer, following Richard Ryder, has called such discrimination against nonhuman animals "speciesism"—"a prejudice or attitude of bias toward the interests of members of one's own species and against those of members of other species."[39]

The implication of Singer's argument is that, where practicable, we must avoid inflicting any suffering on sentient beings. Accordingly, supporters of animal liberation advocate the prohibition of the hunting and slaughtering of all sentient beings (the corollary of which is vegetarianism), the prohibition of vivisection, and the prohibition of "factory farming." Although Singer's major focus has been the abuse of domestic animals, his argument also provides a justification for the protection of the habitat of wild animals, fish, birds, and other sentient fauna. That is, forests and wetlands ought to be protected where it can be shown that they are instrumentally valuable to sentient beings for their "comfort and well-being" in providing nesting sites, breeding habitat, and sustenance.

The attractiveness of Singer's method of argument is that it employs a familiar principle that is widely accepted (i.e., that pleasure is good and pain is bad) and then proceeds to logically press this rationale in such a way that those who accept the premise are forced to accept his conclusion. Moreover, the analogy with racism is used to underscore the point that the animal liberation movement is but the latest in a series of humanitarian or emancipatory movements that began with the antislavery campaigns and later broadened to include the anticolonial and women's movements, all of which have sought to expose and eradicate discriminatory practices on behalf of oppressed groups. In this respect, it is presented as part of a praiseworthy trend of moral and political progress—as one more step along the path toward universal justice, or, as the environmental ethicist J. Baird Callicott has described it, "the next and most daring development of political liberalism."[40]

Ecocentric philosophers, however, have been critical of Singer's moral philosophy for regarding *nonsentient* beings as morally inconsequential. As Rodman has observed, Singer's philosophy leaves the rest of nature

> in a state of thinghood, having no intrinsic worth, acquiring instrumental value only as resources for the well-being of an elite of sentient beings. Homocentrist rationalism has widened out into a kind of zoocentrist sentientism.[41]

Trees, for example, are considered to be valuable only insofar as they provide habitat, can be turned into furniture, or otherwise rendered serviceable to the needs of sentient life-forms. To the extent that synthetic substitutes can be made to perform the services of nonsentient life-forms, then the latter will be rendered dispensible.

Some environmental philosophers have also mounted a more subtle critique of the animal liberation perspective. According to John Rodman, not only does this approach render *non*sentient beings morally inconsequential but it also subtly degrades *sentient non*human beings by regarding them as analogous to "defective" humans who likewise cannot fulfil any moral duties.[42] Rodman sees this tendency to regard sentient nonhumans as having the same standing as defective or inferior human beings as analogous to (and as ridiculous as) dolphins regarding humans "as defective sea mammals who lack sonar capability."[43] The result is that the unique modes of existence and special capabilities of these nonhuman beings are overlooked.

A further criticism levelled against Singer's moral philosophy is that it is atomistic and therefore unsuitable for dealing with the complexities of environmental problems, which demand an understanding and recognition of not only whole species but also the interrelationships between different natural cycles, systems, and populations. According to Rodman, the progressive extension model of ethics (which includes Christopher Stone's argument for the legal protection of the rights of nonsentient entities, discussed in the following section) tends

> to perpetuate the atomistic metaphysics that is so deeply imbedded in modern culture, locating intrinsic value only or primarily in individual persons, animals, plants, etc., rather than in communities or ecosystems, since individuals are our paradigmatic entities for thinking, being conscious, and feeling pain.[44]

Finally, critics have pointed to the tension between Singerian justice and an ecological perspective by noting that animal liberation, when pressed to its logical conclusion, would be obliged to convert all nonhuman animal carnivores to vegetarians, or, at the very least, replace predation in the food chain with some kind of "humane" alternative that protects, or at least minimizes the suffering of, sentient prey. As Fox argues, besides representing "ecological lunacy," animal liberation

> would serve, in effect, to endorse the modern project of totally domesticating the nonhuman world. Moreover, it would also condemn as immoral those "primitive" cultures in which hunting is an important aspect of existence.[45]

Singer has in fact admitted that the existence of nonhuman carnivores poses a problem for the ethics of animal liberation. Despite this concession, he nonetheless counsels a modification to the dietary habits of at least some domestic animals in referring his readers to recipes for a vegetarian menu for their pets![46]

To conclude, then, animal liberation has mounted a compelling challenge to anthropocentrism in pointing to its many logical inconsistencies. However, Singer's alternative criterion of moral considerability (i.e., sentience), while a *relevant* and significant factor, is too limited and not sufficiently ecologically informed to provide the exclusive criterion of a comprehensive environmental ethics. As we shall see, ecocentric theorists have identified broader, less "human analogous" and more ecologically relevant criteria to determine whether a being or entity has "interests" deserving of moral consideration.

Ecocentrism

Ecocentric environmentalism may be seen as a more wide-ranging and more ecologically informed variant of preservationism that builds on the insights of the other streams of environmentalism thus far considered. Whereas the early preservationists were primarily concerned to protect wilderness as sublime scenery and were motivated mainly by aesthetic and spiritual considerations, ecocentric environmentalists are also concerned to protect threatened populations, species, habitats, and ecosystems *wherever situated* and irrespective of their use value or importance to humans. (This kind of concern is well illustrated by the activities of the international environmental organization Greenpeace.) In particular, ecocentric environmentalists strongly support the preservation of large tracts of wilderness as the best means of enabling the flourishing of a diverse nonhuman world. Accordingly, in what I refer to as New World regions such as North America, Australia, and New Zealand (where significant areas of wilderness still remain), it is not surprising that the greatest concentration of ecocentric activists can usually be found in organizations, campaigns, or movements that promote the protection of wilderness. Two noteworthy examples here are the Earth First! movement in the United States and The Wilderness Society in Australia.

Much of the basic outline of an ecocentric perspective has already been foreshadowed in chapter 1 and in the criticisms made of resource conservation, human welfare ecology, preservationism, and animal liberation. An ecocentric perspective may be defended as offering a more encompassing approach than any of those so far examined in that it (i) recognizes the full range of human interests in the nonhuman world (i.e., it incorporates yet goes beyond the resource conservation and human welfare ecology perspectives); (ii) recognizes the interests of the nonhuman community (yet goes beyond the early preservationist perspective); (iii) recognizes the interests of future generations of humans and nonhumans; and (iv) adopts a holistic rather than an atomistic perspective (contra the animal liberation perspective) insofar as it values populations, species, ecosystems, and the ecosphere *as well as* individual organisms.[47]

Now defenders of an animal liberation perspective might argue that their perspective is quite adequate to secure the protection of many nonsentient entities, such as ecosystems, and that for all *practical* purposes it is as good as an ecocentric perspective. This is because, as we saw in the previous section, if we attribute intrinsic value to all sentient beings, then we must also protect whatever is instrumentally valuable to *them* (e.g., *their* habitats and food sources). This would provide a case for the protection of forests, wetlands, and any other habitat upon which sentient nonhuman beings depend for their survival and wellbeing. However, as we also saw in the previous section, ecocentric theorists have argued that this kind of approach not only leaves the rest of nature in a state of "thinghood"—the only purpose of which is to service an elite of sentient beings—but that it is also too atomistic and, therefore, "unecological" in the way in which it distributes intrinsic value in the world. For example, this kind of approach would attribute equal intrinsic value and, hence, equal moral consideration to the individual members of a native species or an endangered species as it would to the individual members of an introduced species or an abundant species (assuming the degree of sentience of each species to be roughly equivalent). This approach to the distribution of intrinsic value means that it would be considered no worse to kill, say, twenty members of a native species than it would be to kill ("weed out") twenty members of an introduced, feral species; for the same reason, it would be considered no worse to kill the last twenty members of a sentient endangered species than it would be to kill twenty members of an equally sentient species that exists in plague proportions. Similarly, an animal liberation perspective would attribute the same value to the individual animals that inhabit a flourishing, wild ecosystem as the equivalent number of domesticated or captive wild animals that might be managed by humans on a farm or in a zoo.

Even if one extends intrinsic value to all living organisms (i.e., animals, plants, and microorganisms) the same general kinds of problems apply. This is because such an approach still remains atomistic (i.e., it attributes intrinsic value only to *individual* living organisms) and therefore does not extend any moral recognition to populations, species, ecosystems, and the ecosphere considered as entities in their own right. (I do not discuss a specifically "life-based" stream of environmentalism in this survey of the major streams of environmentalism for the simple reason that, sociologically and politically speaking, this approach does not represent a major stream of environmentalism. Environmentalists who have moved beyond anthropocentrism tend, on the whole, to gravitate toward either the animal liberation approach or a straight-out ecocentric approach.)

Ecocentric theorists are concerned to develop an ecologically informed approach that is able to value (for their own sake) not just individual living organisms but also ecological entities at different levels of aggregation, such as populations, species, ecosystems, and the ecosphere (or Gaia). What would such an ecocentric theory look like? In the following chapter, we shall see that there are at least three different theoretical frameworks that overcome the problem of atomism.

CHAPTER 3

Ecocentrism Explained and Defended

INTRODUCTION

So far, I have been concerned to distinguish a general ecocentric approach from the other major streams of modern environmentalism indentified in chapter 2, namely, resource conservation, human welfare ecology, preservationism, and animal liberation. It still remains to explore the theoretical framework of ecocentrism in a little more detail and address some of the common criticisms and misunderstandings that are often levelled against, or associated with, a general ecocentric perspective. I will also use this opportunity to compare and discuss three distinctive types of ecocentrism within the Western tradition: autopoietic intrinsic value theory, transpersonal ecology, and ecofeminism.

ECOCENTRISM EXPLAINED

Ecocentrism is based on an ecologically informed philosophy of *internal relatedness,* according to which all organisms are not simply interrelated with their environment but also *constituted* by those very environmental interrelationships.[1] According to Birch and Cobb, it is more accurate to think of the world in terms of "events" or "societies of events" rather than "substances":

> Events are primary, and substantial objects are to be viewed as enduring patterns among changing events.... The ecological model is a model of internal relations. No event first occurs and then relates to its world. The event is a synthesis of relations to other events.[2]

According to this picture of reality, the world is an intrinsically dynamic, interconnected web of relations in which there are no absolutely discrete entities and no absolute dividing lines between the living and the nonliving, the animate and the inanimate, or the human and the nonhuman. This model of reality undermines anthropocentrism insofar as whatever faculty we

49

choose to underscore our own uniqueness or specialness as the basis of our moral superiority (e.g., rationality, language, or our tool-making capability), we will invariably find either that there are some humans who do *not* possess such a faculty or that there are some nonhumans who *do*.[3] Nonanthropocentric ethical theorists have used this absence of any rigid, absolute dividing line between humans and nonhumans to point out the logical inconsistency of conventional anthropocentric ethical and political theory that purports to justify the exclusive moral considerability of humans on the basis of our separateness from, say, the rest of the animal world. Indeed, we saw in the previous chapter how Singer used this kind of argument to criticize human-centered ethical theory and defend animal liberation. While there are undoubtedly many important differences in *degree* (as distinct from kind) between all or some humans and nonhumans, as Fox points out, this cuts both ways; for example, there are countless things that other animals do better than us.[4] (And there are also innumerable differences in capacities that separate nonhuman life-forms from each another.) From an ecocentric perspective, to single out only *our* special attributes as the basis of our exclusive moral considerability is simply human chauvinism that conveniently fails to recognize the special attributes of other life-forms: it assumes that what is distinctive about humans is *more worthy* than, rather than simply *different* from, the distinctive features of other life-forms.[5] John Rodman has called this the "differential imperative," that is, the selection of what humans do best (as compared to other species) as the measure of human virtue and superiority over other species. Rodman traces this idea in Western thought as far back as Socrates, who saw the most virtuous human as "the one who most fully transcends their animal and vegetative nature."[6] The upshot, of course, is that one becomes a better human if one reinforces the differential imperative by maximizing one's "species-specific differentia." (Moreover, as Benton points out, the putative human/animal opposition may sometimes be seen as serving "as a convenient symbolic device whereby we have attributed to animals the dispositions we have not been able to contemplate in ourselves."[7])

Ecocentric theorists have also pointed out how new scientific discoveries have served to challenge long standing anthropocentric prejudices. As the Copernican and Darwinian revolutions have shown, scientific discoveries can have a dramatic impact on popular conceptions of, and orientations toward, nature. This is not to argue that science can or ought to determine ethics or politics but merely to acknowledge that in modern times the credibility of any Western philosophical worldview is seriously compromised if it is not at least cognizant of, and broadly consistent with, current scientific knowledge. It is indeed ironic that while an ecocentric orientation is often wrongly criticized for resting on an "anti-science," mystical idealization of nature, many proponents of ecocentrism are quick to point out that the philosophical premises of ecocentrism (i.e., the model of internal relations) are actually *more* consistent with modern science than the premises of anthropocentrism, which posit humans as either separate from and above the rest of nature (or if not separate from the rest of nature then nonetheless the acme of evolution). In this respect, ecocentric theorists, far from being anti-science, often *enlist* science to help undermine deeply ingrained anthropocentric assumptions that have found their way into many branches of the social sciences and humanities, including modern political theory. As George Sessions has argued, modern science has "been the single most decisive non-anthropocentric intellectual force in the Western world."[8] Indeed, it has been the mechanistic, materialistic worldview of the Enlightenment (which has shaped so much modern political theory) that has most come under challenge by these new scientific discoveries. Just as the Copernican and Darwinian revolutions helped to undermine the Judeo-Christian, medieval worldview of the "great chain of being" (according to which all life-forms were fixed in a static hierarchy with humans standing above the beasts and below the angels), the picture of ecological and subatomic reality that has emerged from new discoveries in biology and physics has now made inroads into many of the assumptions of the Newtonian worldview.[9] The most pervasive of these are *technological optimism*—the confident belief that with further scientific research we can rationally manage (i.e., predict, manipulate, and control) all the negative unintended consequences of large-scale human interventions in nature; *atomism*—the idea that nature is made up of discrete building blocks and that the observer is therefore completely separate from the observed; and *anthropocentrism*—the belief that there is a clear and morally relevant dividing line between humankind and the rest of nature, that humankind is the only or principal source of value and meaning in the world, and that nonhuman nature is there for no other purpose but to serve humankind.[10]

Clearly, ecocentric theorists are not against science or technology per se; rather they are against scientism (i.e., the conviction that empiric-analytic science is the only valid way of knowing) and technocentrism (i.e., anthropocentric technological optimism). The distinction is crucial. Indeed, many ecocentric theorists are keenly interested in the history and philosophy of science and are fond of pointing out the reciprocal interplay between dominant images of nature (whether derived from science, philosophy, or religion) and dominant images of society.[11] This mutual reinforcement is reflected in the resonance between medieval Christian cosmology and the medieval political order (both of which emphasized a hierarchy of being) and between the Newtonian worldview and the rise of modern liberal democracy (both of which emphasized atomism). Ecocentric theorists are now drawing attention to what Fox has referred to as the "structural similarity" between the ecological model of internal relatedness and the picture of reality that has emerged in modern biology and physics, although it is too early to say what the societal implications of these developments might be.[12] Unlike Capra, I see nothing *inevitable* about the possibility of a new, ecologically informed cultural transformation, although there are certainly many exciting possibilities "in the wind."[13]

The structural similarity between the ecological model of internal relatedness that informs ecocentrism and the picture of reality delivered to us by certain branches of modern science is, of course, no substitute for an ethical and political justification of an ecocentric perspective (although it does serve to undermine the opposing perspective of anthropocentrism). As I noted earlier, in modern times general consistency with science is merely a necessary—as distinct from a sufficient—condition for the acceptance of an alternative philosophical worldview in the West. In this respect, I agree with Michael Zimmerman's observations concerning the relevance of science to environmental ethics and politics: that it may help to inspire and prepare the ground for a new orientation toward nature and "give humanity prudential reasons for treating the biosphere with more care" but that "a change in scientific understanding alone cannot produce the needed change of consciousness."[14] It is no argument, then, simply to appeal to the authority of nature as a justification for a particular political worldview. It is, on the other hand, perfectly reasonable to question an opposing worldview on the ground that the assumptions on which it is based have been shown by science to be erroneous.

The ecocentric recognition of the interrelatedness of all phenomena together with its prima facie orientation of inclusiveness of all beings means that it is far more protective of the Earth's life-support system than an anthropocentric perspective. As Michael Zimmerman has argued in addressing the practical consequences of an anthropocentric perspective:

> If humankind is understood as the goal of history, the source of all value, the pinnacle of evolution, and so forth, then it is not difficult for humans to justify the plundering of the natural world, which is not human and therefore "valueless."[15]

When anthropocentric assumptions of this kind are combined with a powerful technology, the capacity for environmental destruction increases dramatically.

Anthropocentrism of this extreme kind may be seen as a kind of ecological myopia or unenlightened self-interest that is blind to the ecological circularities between the self and the external world, with the result that it leads to the perpetuation of unintended and unforeseen ecological damage. An ecocentric perspective, in contrast, recognizes that nature is not only more complex than we presently know but also quite possibly more complex, in principle, than we *can* know—an insight that has been borne out in the rapidly expanding field of chaos theory.[16]

Although the anthropocentric resource conservation and human welfare ecology streams of environmentalism adopt a general ethic of prudence and caution based on an ecologically enlightened self-interest, they differ from an ecocentric perspective in that they see the ecological tragedy as essentially a *human* one. Those belonging to the ecocentric stream, on the other hand, see the tragedy as *both* human and nonhuman. This is because a thoroughgoing ecocentric perspective is one that, "within obvious kinds of practical limits, allows all entities (including humans) *the freedom to unfold in their own way unhindered by the various forms of human domination.*"[17] Such a general perspective may be seen as seeking "emancipation writ large." In according ontological primacy to the internal relatedness of all phenomena, an ecocentric perspective adopts an "existential attitude of mutuality" in recognition of the fact that one's personal fulfilment is inextricably tied up with that of others.[18] This is not seen as a resignation or self sacrifice but rather as a *positive affirmation* of the fact of our embeddedness in ecological relationships.

The ecological model of internal relatedness upon which ecocentrism rests applies not only in respect of human-nonhuman relations but also in respect of relations among humans: in a biological, psychological, and social sense we are all constituted by our interrelationships between other humans, and our political, economic, and cultural institutions. As Birch and Cobb emphasize, we do not exist as separate entities and *then* enter into these relations. From the moment we are born, we are constituted by, and coevolve within the context of, such relations.[19] According to this model, we are neither completely passive and determined beings (as crude behaviorists would have it) nor completely autonomous and self-determining beings (as some existentialists would have it). Rather, we are *relatively* autonomous beings who, by our purposive thought and action, help to constitute the very relations that determine who we are.[20] Of course, this kind of social interactionist model is hardly new to the social sciences. For example, in social psychology it is found in the theories of symbolic interactionism and phenomenology. In political philosophy a similar social model is implicit in the many communitarian and socialist political philosophies that seek the mutual self-realization of all in preference to the individual self-realization of some. This helps to explain why there is a much greater elective affinity—and hence a much greater potential for theoretical synthesis—between ecocentrism and communitarian and socialist political philosophies than there is between ecocentrism and individualistic political philosophies such as liberalism, as we saw in chapter 1. Ecocentric political theorists have generally discarded what Callicott has aptly described as "the threadbare metaphysical cloth from which classical utilitarianism [and, I would add, Lockean liberalism] is cut." This is because, as Callicott puts it,

> Utilitarianism [indeed liberalism in general] assumes a radical individualism or rank social atomism completely at odds with the relational sense of self that is consistent with a more fully informed

evolutionary and ecological understanding of terrestrial and human nature.[21]

It should be clear, however, that ecocentric theorists are not seeking to discard the central value of autonomy in Western political thought and replace it with something completely new. Rather, ecocentric theorists are merely concerned to revise the notion of autonomy and incorporate it into a broader, ecological framework.

The word autonomy is derived from the Greek *autos* (self) and *nomos* (law) and means, literally, to live by one's own laws. This is similar to Immanuel Kant's influential formulation according to which an autonomous person is someone who acts from self-imposed principle (as distinct from personal whim or externally imposed commands). Ecocentric theorists have carried forward this basic notion of autonomy as self-determination. However, they have extended the interpretation and application of the notion by radically revising the notion of "self." After all, if we take autonomy to mean self-determination, this still begs the question as to what kind of "self" we are addressing. In lieu of the atomistic and individualistic self of liberalism or the more social self of socialism, ecocentric theorists have introduced a broader, ecological notion of self that incorporates these individual and social aspects in a more encompassing framework. From the perspective of the ecological model of internal relations, the liberal idea of autonomy as independence from (or "freedom from") others is seen as philosophically misguided. (To the extent that interconnectedness with others is acknowledged under this particular liberal interpretation, it is likely to be experienced as threatening, as causing a loss of self). While socialists tend to adopt a more relational model of self (which sometimes encompasses our relations with the nonhuman world), this still remains embedded in an anthropocentric framework.

Of course, the ecocentric reformulation of autonomy does not mean that the boundary between one's individual self and others completely falls away. Rather, the reformulation merely seeks to emphasize the flexible or soft nature of the boundaries between self and other (which is why ecocentric theorists often refer to the individual self as forming part of a "larger self"). Evelyn Fox Keller encapsulates the permeable nature of the boundaries between self and other in her concept of "dynamic autonomy." As Keller explains, this dynamic concept of autonomy "is a product at least as much of relatedness as it is of delineation; neither is prior."[22] From an ecocentric perspective, the exercise of dynamic autonomy requires psychological maturity and involves a sensitive mediation between one's individual self and the larger whole. This does not mean having control over others but rather means having *a sense of competent agency in the world* in the context of an experience of continuity with others. In contrast, the quest for radical independence from others, or power over others, leads to an objectification of others, and a denial of their own modes of relative autonomy or subjectivity.

What is new about an ecocentric perspective (and Keller shares this perspective) is that it extends the notion of autonomy (and the interactionist model of internal relations on which it is based) to a broader and more encompassing pattern of layered interrelationships that extend beyond personal and societal relations to include relations with the rest of the biotic community. This means that the nonhuman world is no longer posited simply as the background or means to the self-determination of individuals or political communities, as is the case in most modern political theorizing. Rather, the different members of the nonhuman community are also appreciated as important in their own terms, as having their own (varying degrees of) relative autonomy and their own modes of being. The implications of applying this expanded model of internal relations to social and political thought are far reaching. As Zimmerman has put it, "the paradigm of internal relations lets us view ourselves as manifestations of a complex universe; we are not apart but are moments in the openended, novelty-producing process of cosmic evolution."[23]

SOME COMMON CRITICISMS AND MISUNDERSTANDINGS

Ecocentrism's challenge to cultural and political orthodoxy has been widely resisted and misunderstood by critics for a variety of reasons: that it is impossible, misanthropic (or at least insulting to some humans, notably the oppressed), impractical, and/or based on an all too convenient idealization of nature. Some resistance is, of course, to be expected of a perspective that, as George Sessions has put it, is mounting a philosophical challenge to "the pervasive metaphysical and ethical anthropocentrism that has dominated Western culture with classical Greek humanism and the Judeo-Christian tradition since its inception."[24] But is such resistance warranted? In the remainder of this section I address five common objections that have contributed to this resistance to ecocentrism.

One common criticism is that it is impossible to perceive the world *other* than from an anthropocentric perspective since we are, after all, *human* subjects. This criticism, however, entirely misses the point of the critique of anthropocentrism by conflating the identity of the perceiving subject with the content of what is perceived and valued, a conflation that Fox has called the "anthropocentric fallacy."[25] In particular, this kind of understanding conflates the trivial and tautological sense of the term anthropocentrism (i.e., that we can only ever perceive the world as human subjects—who can argue against this?) and the substantive and informative sense of the term (the unwarranted, differential treatment of other beings on the basis that they do not belong to our *own* species).[26] Ecocentric theorists are not claiming that we must, or indeed can, know *exactly* what it is like to be, say, a kangaroo (although there are meditation traditions and forms of shamanic journeying that enable humans to experience the world as other beings).[27] As Barbara Noske explains, "there is a sense in which we cannot know the Other (whether it be other species, other cultures, the other sex or even each other) [however] we must remind ourselves that other meanings exist, even if we may be severely limited in our understanding of them."[28] As Fox points out, to say that humans cannot be nonanthropocentric is like saying that a male cannot be nonsexist or that a white person cannot be nonracist because they can only perceive the world as male or white subjects.[29] This understanding ignores the fact that males and whites are quite capable of cultivating a non-sexist or nonracist consciousness or, in this case, that humans are quite capable of cultivating a nonanthropocentric consciousness.

A second misconception of ecocentrism is to interpret its sustained critique of anthropocentrism as anti-human and/or as displaying an insensitivity to the needs of the poor and the oppressed by collectively blaming the human species as a whole for the ecological crisis (rather than singling out specific nations, groups, or classes). However, this criticism fails to appreciate the clear distinction between a *non*anthropocentric and a *mis*anthropic perspective.[30] Ecocentrism is not against humans per se or the celebration of humanity's special forms of excellence; rather, it is against the ideology of human chauvinism. Ecocentric theorists see each human individual and each human culture as just as entitled to live and blossom as any other species, *provided* they do so in a way that is sensitive to the needs of other human individuals, communities, and cultures, and other life-forms generally. Moreover, many critics of ecocentrism fail to realize that a perspective that seeks emancipation writ large is one that *necessarily* supports social justice in the human community. Given that it is patently the case that not all humans are implicated in ecological destruction to the same degree, then it follows that ecocentric theorists would not expect the costs of environmental reform to be borne equally by all classes and nations, regardless of relative wealth or privilege. That many ecocentric theorists have given special theoretical attention to human-nonhuman relations arises from the fact that these relations are so often neglected by theorists in the humanities and social sciences. It does not arise from any lack of concern or lack of theoretical inclusiveness with regard to human emancipatory struggles.

Before leaving this point, it should be noted that some ecophilosophically minded writers (e.g., David Ehrenfeld in *The Arrogance of Humanism*) have been critical of humanism in general, rather than just anthropocentrism. This can be misleading, however, since humanism does not represent one single idea, such as human self-importance or the celebration of humanity as the sole and sufficient source of value and inspiration in the world, although these have been central ideas in humanism and are the main bone of contention of nonanthropocentric ecophilosophers.[31] Rather, humanism is a complex tapestry of ideas, many strands of which are anthropocentric, yet some strands of which remain worthwhile and consistent with an ecocentric perspective. As Blackham explains, "the 'open mind,' the 'open society,' and the sciences and 'humanities' are the glory of humanism and at the same time a widely shared inheritance."[32] In view of this, it is more to the point simply to criticize the many anthropocentric assumptions embedded in our humanist heritage rather than to equate anthropocentrism with humanism and thereby condemn humanism in its entirety.

A third criticism is that ecocentrism is a passive and quietistic perspective that regards humans as no more valuable than, say, ants or the AIDS virus. However, ecocentrism merely seeks to cultivate a prima facie orientation of nonfavoritism; it does not mean that humans cannot eat or act to defend themselves or others (including other threatened species) from danger or life-threatening diseases.[33] In this respect, the degree of sentience of an organism and its degree of self-consciousness and capacity for richness of experience are relevant factors (as distinct from exclusive criteria) in any ethical choice situation alongside other factors, such as whether a particular species is endangered or whether a particular population is crucial to the maintenance of a particular ecosystem.[34] A nonanthropocentric perspective is one that ensures that the interests of nonhuman species and ecological communities (of varying levels of aggregation) are not ignored in human decision making *simply* because they are not human or because they are not of instrumental value to humans. It does not follow from this prima facie orientation of nonfavoritism, however, that the actual outcome of human decision making must necessarily favor noninterference with other life-forms. Humans are just as entitled to live and blossom as any other species, and this inevitably necessitates some killing of, suffering by, and interference with, the lives and habitats of other species.[35] When faced with a choice, however, those who adopt an ecocentric perspective will seek to choose the course that will minimize such harm and maximize the opportunity of the widest range of organisms and communities—*including ourselves*—to flourish in their/our own way. This is encapsulated in the popular slogan "live simply so that others [both human and nonhuman] may simply live."[36]

A fourth criticism against ecocentrism is that it is difficult to translate into social, political, and legal practice. How, many sceptics ask, can we ascribe rights to nonhumans when they cannot reciprocate? My primary answer to this kind of criticism is that it is neither necessary nor ultimately desirable that we ascribe legal rights to nonhuman entities to ensure their protection. However, it also needs to be pointed out that there is no a priori reason why legal rights cannot be ascribed to nonhuman entities. As Christopher Stone has argued, the idea of conferring legal rights on nonhumans is not "unthinkable" when it is remembered that legal rights are conferred on "nonspeaking" persons such as infants and fetuses, on legal fictions such as corporations, municipalities and trusts, and on entities such as churches and nation states.[37] Given that there is no common thread or principle running through this anomalous class of right holders, Stone argues that there is no good reason *against* extending legal rights to natural entities. Stone proposes that the rights of nonhuman entities (or, in his language, "natural objects") be defended in the same way as "human vegetables," that is, by the appointment of a Guardian or Friend who would ensure that the natural entity's interests were protected (e.g., by administering a trust fund and instigating legal actions on its behalf in order to make good any injury inflicted on it.) Stone's proposal may be seen as an even more daring adventure in liberalism than animal liberation insofar as it seeks to provide the means of legally protecting the special interests of nonhuman and nonsentient entities such as forests, rivers, and oceans.

While Stone's proposals may serve an important educative and protective purpose in respect of nonhuman interests, there is nonetheless an element of absurdity in the notion of extending rights to nonhumans on the basis of a *contractarian* notion of rights, whereby a right must be accompanied by a correlative duty. Stone appears to lean toward such a view in his suggestion that the trust funds established for the benefit of a natural entity might also be used to satisfy judgments *against* that entity (e.g., a river might be liable for the damage inflicted by its flooding and destroying crops!) although he admits that such an idea would prove to be troublesome. As Stone asks: "When the Nile overflows, is it the 'responsibility' of the river? the mountains? the snow? the hydrological cycle?"[38] Stone also canvasses the possibility of "an electoral apportionment that made some systematic effort to allow for the representative 'rights' of nonhuman life."[39] Of course, the first kind of scenario could be avoided by employing a noncontractarian theory of rights (i.e., as not necessarily entailing reciprocal duties), yet there is still something strained and ungainly in the attempt to extend to the nonhuman world political concepts that have been especially tailored over many centuries to protect *human* interests. This highlights the need to search for simpler and more elegant ways of enabling the flourishing of a rich and diverse nonhuman world without resorting to the extension to the nonhuman realm of peculiarly *human* political and legal models such as justice, equality, and rights.[40] As John Livingston points out, extending liberal egalitarian ideals in this way "anthropomorphizes the nonhuman world in order to include it in a human ethical code."[41] Similarly, John Rodman has suggested that the "liberation of nature" requires not the extension of human-like rights to nonhumans but the liberation of the nonhuman world from "the status of human resource, human product, human caricature."[42] It is indeed noteworthy that one of the doyens of modern liberal theory—John Rawls—in discussing the limits of his liberal theory of justice, has stated in passing that it does not seem possible to extend the contract doctrine so as to include them [i.e, creatures lacking a capacity for a sense of justice] in a natural way. A correct conception of our relations to animals and to nature would seem to depend [instead] upon a theory of the natural order and our place in it.[43]

The above reservations concerning the appropriateness of extending legal rights to nonhumans are hardly fatal to ecocentrism; nor do they provide any argument for resorting to anthropocentrism through want of appropriate legal mechanisms. Rather, they emphasize the importance of a general change in consciousness and suggest that a gradual cultural, educational, and social revolution involving a reorientation of our sense of place in the evolutionary drama is likely to provide a better long term protection of the interests of the nonhuman world than a more limited legal revolution of the kind envisaged by Stone. In the short term, the abo1 ve reservations concerning the applicability of liberal categories to the nonuman world highlight the need for us to rethink the ways in which we might legally protect the interests of the nonhuman world. Indeed, there are already existing alternative legislative precedents that avoid the language of rights but nonetheless ensure that government departments and courts consider both human *and* nonhuman interests when administering environmental legislation or adjudicating land-use conflicts.[44]

Finally, some critics are cynical of ecocentrism because they consider that it interprets nature selectively as something that is essentially harmonious, kindly, and benign (ignoring suffering, unpredictability, and change), thus providing an all too convenient model for human relations. Alternatively, critics have argued that the popular ecological views of some Green thinkers lean toward an idealization of nature or employ outmoded ecological notions (such as the "balance of nature") that have little to do with the way nature in fact operates.[45] My response to these criticisms is that ecocentric theorists simply do not need to depict nature as having a kindly human face or to show that nature is essentially benevolent or benign in order that humans respect it and regard it as worthy. If we try to judge the nonhuman world by human standards as to what is "kindly," we will invariably find it wanting.[46] Nonhuman nature knows no human ethics, it simply *is*.

In any event, appealing to the authority of nature (as known by ecology) is no substitute for ethical argument.[47] Ecological science cannot perform the task of normative justification in respect of an ecocentric political theory because it does not tell us why we *ought* to orient ourselves toward the world in a particular way. It can inform, inspire, and redirect our ethical and political theorizing, but it cannot justify it. *That* is the task of ethical and political theory. However, a general familiarity with new developments in science is important to an ecocentric perspective (the employment of outmoded concepts of nature *does* serve to detract from the force and credibility of ecopolitical argument). As we have already seen, a general familiarity with new developments in science by social and political theorists can enhance our understanding of the world around us, improve the general grounding and credibility of a political theory, and provide the basis for challenging opposing worldviews on the grounds that the assumptions on which they are based have been shown to be erroneous.

THREE VARIETIES OF ECOCENTRISM

Having explained and defended a general ecocentric perspective, it now remains to explore some particular theoretical and cultural expressions of ecocentrism. Indeed, there is a wide range of different approaches that are consistent with a general ecocentric perspective. Examples include axiological (i.e., value theory) approaches that argue for the intrinsic value of all living entities *as well* as such "systemic" entities as populations, species, ecosystems, and the ecosphere; the psychological-cosmological approach that is being developed under the name of "deep ecology" or, more recently, "transpersonal ecology"; the antihierarchical and personal ethic of care and reciprocity defended by ecofeminism; certain Eastern philosophies such as Taoism and Buddhism that emphasize the interconnectedness of all phenomena and the importance of humility and compassion; and the animistic cosmologies of many indigenous peoples who see and respect the nonhuman world as alive and enspirited.

I will explore three complementary expressions of ecocentrism within the Western tradition: autopoietic intrinsic value theory (which I see as an improvement on the popular, ecosystem-based "land ethic" of Aldo Leopold), transpersonal ecology, and ecofeminism.[48] In view of the increasing influence and popularity of both deep/transpersonal ecology and ecofeminism within Green circles, particular attention will given to clarifying the areas of overlap and difference between these two distinctive ecocentric approaches.

Autopoietic Intrinsic Value Theory

Autopoietic intrinsic value theory—outlined by Fox under the name of "autopoietic ethics"—represents one kind of intrinsic value theory approach that is capable of providing a sound theoretical basis for ecocentrism. An autopoietic approach attributes intrinsic value to all entities that display the property of *autopoiesis,* which means "self-production" or "self-renewal" (from the Greek *autos,* "self," and *poiein,* "to produce").[49] Autopoietic entities are entities that are "primarily and continuously concerned with the regeneration of their own organizational activity and structure."[50] It is precisely this characteristic of self-production or self-renewal (as distinct from merely self-organization) that distinguishes living entities from self-correct-ing machines that appear to operate in a purposive manner (such as guided missiles). In other words, the primary product of the operations of living systems, as distinct from mechanical systems, is themselves, not some goal or task external to themselves.[51] Autopoietic entities are therefore *ends in themselves,* which, as Fox points out, "amounts to the classical formulation of intrinsic value."[52] This means that autopoietic entities (e.g., populations, gene pools, ecosystems, and individual living organisms) are deserving of moral consideration in their own right.

An autopoietic approach provides a sounder theoretical basis for ecocentrism than the ethical holism of Aldo Leopold's famous land ethic, which declares that "A thing is right when it tends to preserve the integrity, stability, and beauty of the biotic community. It is wrong when it does otherwise."[53] The problem with this ethic, as animal liberation proponents point out, is that it is vulnerable to the charge of "environmental fascism" in that it provides no recognition of the value of *individual* organisms. This is because, considered on its own, it can be interpreted as suggesting that individuals are dispensible—indeed, might need to be sacrificed for the good of the whole.[54]

An autopoietic approach to intrinsic value is not vulnerable to the objections that are associated with either extreme atomism or extreme holism. Whereas atomistic approaches attribute intrinsic value only to individual organisms, and whereas an unqualified holistic approach attributes intrinsic value only to whole ecosystems (or perhaps only the biosphere or ecosphere itself), an autopoietic approach recognizes the value of *all* process-structures that "continuously strive to produce and sustain their own organizational activity and structure."[55] That is, an autopoietic approach recognizes the value not only of individual organisms but also of species, ecosystems, and the ecosphere ("Gaia").

Transpersonal Ecology

In contrast to the autopoietic approach, which proceeds via an axiological (i.e., value theory) route, transpersonal ecology proceeds by way of a cosmological and psychological route and is concerned to address the way in which we understand and experience the world.[56] The primary concern of transpersonal ecology is the cultivation of a wider sense of self through the common or everyday psychological process of *identification* with others. This should not be understood simply as a reformulation of the Kantian Categorical Imperative (i.e., dutiful altruism) or the Golden Rule (i.e., do unto others as you would have others do unto you) since these remain axiological approaches that are concerned with moral *obligations* that may or may not correspond with one's personal (or "heartfelt") inclinations.[57] (In any event, it should be noted here that Kant regarded *only* humans as ends in themselves.[58]) In contrast to axiological approaches, which issue in moral injunctions or a code of conduct (i.e., "you *ought* to respect other beings, regardless of how you might personally experience them"), transpersonal ecology is concerned to find ways in which we may experience a *lived sense* of identification with other beings. Indeed, as Fox points out, transpersonal ecology explicitly rejects approaches that issue in moral injunctions and advances instead an approach that seeks "to *invite* and *inspire* others to realize, in a this-worldly sense, as expansive a sense of self as possible."[59] As Arne Naess explains, if your sense of self embraces other beings, then "you need no moral exhortation to show care" toward those beings.[60] The cultivation of this kind of expansive sense of self means that compassion and empathy naturally flow as part of an individual's way of being in the world rather than as a duty or obligation that must be performed regardless of one's personal inclination.[61]

The transpersonal ecology approach is described as both cosmological *and* psychological because it proceeds from a particular picture of the world or cosmos—that we are, in effect, all "leaves" on an unfolding "tree of life"—to a psychological identification with all phenomena (i.e., with all leaves on the tree). Fox refers to this approach as *transpersonal* ecology because it is concerned to cultivate a sense or experience of self that extends *beyond* one's egoistic, biographical, or personal sense of self to include all beings.[62] This should not be confused with a generalized form of narcissism as this simply involves the projection of one's ego into a larger sphere (i.e., the self that is revered is still confined to the narrow, egoistic, particle-like sense of self). In contrast, transpersonal ecology (which draws on insights from, among other sources, transpersonal psychology and the teachings of Mahatma Gandhi) is addressing a much broader, transpersonal (i.e., trans-egoistic) sense of "ecological self" that embraces other beings (human and nonhuman) and ecological processes. The movement from an atomistic, *ego*-centric sense of self toward an expansive, *eco*centric or transpersonal sense of self is seen as representing a process of psychological maturing.[63] In other words, transpersonal ecology is concerned to expand the circle of human compassion and respect for others beyond one's particular family and friends and beyond the human community to include the entire ecological community. The realization of this expansive, ecocentric sense of self is brought about by the process of identification with other beings (i.e., the cultivation of an empathic orientation toward the world) in recognition of the fact that one's own individual or personal fate is intimately bound up with the fate of others. It is noteworthy that deep/transpersonal ecology theorists provide a sophisticated theoretical articulation of the experience of many environmental activists as revealed in empirical research. As Lester Milbrath found in his sociological survey, "environmentalists, much more than non-environmentalists, have a generalized sense of compassion that extends to other species, to people in remote communities and countries, and to future generations."[64]

Some critics might object that this kind of approach attempts to derive an "ought" from an "is," in that it proceeds from the fact of our interconnectedness with the world to a particular kind of normative orientation toward the world. However, transpersonal ecologists do not seek to argue that the fact of out interconnectedness *logically* implies a caring orientation toward the world (indeed, this fact does not *logically* imply *any* kind of normative orientation). Rather, transpersonal ecologists are offering an "experiential invitation" rather than issuing a moral injunction. As Fox explains:

> For transpersonal ecologists, given a deep enough understanding of the way things are, the response of being inclined to care for the unfolding of the world in all its aspects follows "naturally"—not as a *logical* consequence but as a *psychological* consequence; as an expression of the spontaneous unfolding (development, maturing) of the self.[65]

The autopoietic intrinsic value theory approach and the transpersonal ecology approach each have different advantages and are appropriate in different contexts. For example, the autopoietic approach is more suitable to translation into legal and political practice than the transpersonal ecology approach. After all, it makes sense to enact legislation that demands the recognition of certain intrinsic values, whereas it makes no sense to enact legislation that demands that people identify more widely with the world around them.[66] Indeed, there are already existing legislative precedents that are consistent with an autopoietic intrinsic value theory approach rather that an atomistic intrinsic value approach in that they value for their own sake both individual living organisms as well as entities such as ecosystems.[67]

The transpersonal ecology approach, in contrast, is more appropriately pursued in the community through educational and cultural activities (although these activities can, of course, be encouraged and financially supported by the state). Transpersonal ecology, in other words, lends itself far more to a "bottom-up" rather than a "top-down" approach to social change. Transpersonal ecology may thus be seen as forming part of the vanguard of the cultivation of a new worldview, a new culture and character, and new political horizons that are appropriate to our times. As we saw in chapter 1, this emphasis on cultural renewal and reenvisioning our place in nature forms an essential component of ecocentric emancipatory thought. In this respect, transpersonal ecology may be seen as being more expressive of the cultural aspirations of ecocentric emancipatory thought than an autopoietic intrinsic value theory approach, although the latter is an essential complement. That is, cultural renewal and legislative reform must go hand in hand.

Ecofeminism

Like transpersonal ecology, ecofeminism is concerned with our sense of self and the way in which we experience the world rather than with formal value theory. Like transpersonal ecology, ecofeminism also proceeds from a process oriented, relational image of nature and seeks mutualistic social and ecological relationships based on a recognition of the interconnectedness, interdependence, and diversity of all phenomena.[68]

However, unlike transpersonal ecology, ecofeminism has taken the historical/symbolic association of women with nature as demonstrating a special convergence of interests between feminism and ecology. The convergence is seen to arise, in part, from the fact that patriarchal culture has located women somewhere between men and the rest of nature on a conceptual hierarchy of being (i.e., God, Man, Woman, Nature). This has enabled ecofeminists to identify what they see as a similar logic of domination between the destruction of nonhuman nature and the oppression of women. Indeed, it is a central claim of many ecofeminists (writing mostly, but not exclusively, from a radical feminist perspective) that "the larger culture's devaluation of natural processes was a product of masculine consciousness."[69] As Simone de Beauvoir observed in her wide ranging exploration of the "second sex," women—like nonhuman animals—have usually been more preoccupied with the regeneration and repetition of life, whereas men have usually been free to seek ways of transcending life by remodelling, reshaping, and recreating the future through technology and symbols. Whereas women's activity has usually been perishable, involving "lower level" transformations of nature, men's activity has usually been more lasting, involving major transformations of nature and culture.[70]

Although many feminists have rejected the association of women and nature as burdensome, ecofeminists, while recognizing that the association has been used to oppress women in the past, have nonetheless embraced it as a source of empowerment for women and the basis of a critique of the male domination of women *and* nonhuman nature. This is an explicitly ecofeminist project because it exposes and celebrates what has traditionally been regarded as Other—both woman *and* nonhuman nature—in the context of a far-reaching critique of hierarchical dualism and "masculine" culture. Ecofeminists seek to subvert the dominant valuation of what human characteristics and activities are most valuable. This project entails a rejection of many of the "advances" of patriarchal culture and a celebration of the previously undervalued nurturing characteristics of women.

In contrast to the more secular leanings of socialist and liberal feminists, many ecofeminists are vitally interested in cultivating an ecofeminist spirituality, whether it be through retrieving the insights of nonhierarchical pre-Christian cultures or reviving other Earth-based traditional practices (e.g., celebrating the Goddess-oriented culture of Old Europe, pagan rituals, Gaia, the body, natural cycles, and the experience of connectedness and embodiment in general).[71] In this respect, most ecofeminists would have much sympathy with Gary Snyder's sentiment that "our troubles began with the invention of male deities located off the planet."[72] Indeed, both ecofeminist and deep/transpersonal ecology theorists have been particularly critical of the Judeo-Christian heritage. As the ecofeminist theologian Elizabeth Dodson Gray explains, we need to move toward an "embodied ecospirituality" and re-myth Genesis in a way that honors diversity by moving our culture "to *a creation-based valuing of all parts of nature.*"[73]

There is a great deal in common between the particular ecological sensibility or sense of self defended by ecofeminists and deep/transpersonal ecologists. Indeed, many ecofeminist insights are indistinguishable from the "wider indentification" approach of transpersonal ecology. For example, Elizabeth Dodson Gray has suggested that the root of the modern ecological crisis is that "*we do not understand who we are*"; when we realize that we are intimately connected with the larger whole then "what hurts any part of my larger system will hurt me."[74]

Notwithstanding these important commonalities, ecofeminism tends to diverge from transpersonal ecology in two significant ways (although these tendencies are far from uniform among different ecofeminist writers). The first relates to the *kinds of identification* that are emphasized and the *kinds of self* that identify; the second relates to the kinds of theoretical explanation offered to account for the environmental crisis.

Most ecofeminists emphasize a form of identification with the world that is gender specific and based on, or at least begins with, personal contact and familiarity. Indeed, some ecofeminists have criticized the kind of identification defended by transpersonal ecologists for being abstract, impersonal, and preoccupied with the whole at the expense of the parts.[75] However, these criticisms overlook the fact that the transpersonal ecology approach to identification encompasses the personal by including both the whole *and* the parts in what Fox describes as an "outside-in" rather than "inside-out" approach. (Indeed, many ecofeminists arrive at a similar position to transpersonal ecology, albeit from an "inside-out" rather than "outside-in" route.) As Fox explains, it is an approach that "proceeds from a sense of the cosmos (such as that provided by the image of the tree of life) and works inward to each particular individual's sense of commonality with other entities."[76] Fox argues that cosmologically based identification represents a more impartial, inclusive, and, hence, more egalitarian approach to identification than does a *purely* personally based approach, in that it leads one to identify with all of the human and nonhuman world irrespective of one's personal involvement. Moreover, Fox argues that purely personally based forms of identification can lead to excessive partiality, attachment, possessiveness, and parochialism.[77] This does not mean, however, that transpersonal ecologists wish to deny the significance of personally based identification—indeed, Fox acknowledges that this kind of identification is the easiest and most immediate experience of identification available to humans. Rather, transpersonal ecologists simply seek to locate personally based identification in a wider social and ecological context. Moreover, this kind of identification does not seek to draw specific gender boundaries; rather it addresses a kind of identification that is available to both women and men.[78]

Nonetheless, some ecofeminists have argued that there is something special about women's experience that make women better placed than men to identify with nonhuman beings, ecological processes, and the larger whole. This argument takes two forms (although these are not always clearly differentiated). On the one hand, it is often claimed, or more usually implied, that this ease of identification with the rest of nature arises by virtue of what is unique about women's bodies (e.g., ovulation, menstruation, pregnancy, childbirth, and suckling the young). Here, the special connection between women and nature is usually presented as something that is grounded in women's reproductive and associated nurturing capabilities. On the other hand, it is often claimed that this readiness to identify with the rest of nature arises by virtue of women's oppression. That is, the separate social reality of women that has resulted from the division of labor in patriarchal societies (i.e., between the spheres of reproduction [women] and production [men]) is heralded as the basis of an alternative, nurturing, and more caring morality.[79] I will refer to these two arguments as the "body-based argument" and the "oppression argument" respectively. In both cases, the closer connection between woman and nature (whether biologically based or culturally assigned) is embraced as a source of special insight and empowerment for women.

Depending on how they are formulated, these two arguments carry significant insights that can assist in the task of ecological and social reconstruction. However, these arguments also raise a number of problems if pressed too far. While it cannot be denied that male and female bodily experiences do differ in a number of important respects, it is problematic to suggest that the particular bodily experiences that are unique to women confer on women a *superior* (as distinct from merely special) insight into our relatedness with life.[80] Such an argument effectively introduces a new hierarchical dualism that subtly condemns men to an inferior status (of Otherness) on the ground that men's bodily differences render them incapable of participating in the particular kind of body-based consciousness that is believed to confer on women a keener psychological awareness of ecological connectedness. Yet, as transpersonal ecologists and some ecofeminists show, there is no a priori reason why men *and* women cannot both participate in body based forms of identification (whether personal *or* cosmological).[81] To the extent that bodily experiences may differ between men and women, there is no reason why either should be *socially* elevated as superior to the other. Moreover, the nurturing qualities usually associated with women can for the most part be attributed to the social division of labor and therefore can be made more culturally diffuse through shared parenting (the latter reform is, of course, strongly supported by feminists and ecofeminists). Finally, as Joan Griscom observes,

> simply because women are *able* to bear children does not mean that doing so is *essential* to our nature. Contraception clarifies this distinction: the ability to give birth can now be suppressed, and there are powerful ecological pressures in favor of this. In this context, it is important that biology *not* be our destiny.[82]

The oppression argument—that women are in a better position to critically evaluate ecological practices and envision an alternative society by virtue of their *oppression* rather than their biology per se—provides a more defensible reason for paying special attention to the experiences of women. That women have been less implicated than men in major activities and centers of ecological destruction (e.g., the military, the boardroom, science, and bureaucracy) is itself an excellent reason to hear what women have to say on the subject of ecological reconstruction. Most women do occupy a vantage point of "critical otherness" from which they can offer a different way of looking at the problems of both patriarchy *and* ecological destruction. Of course, the same can be said for many other minority groups and classes such as indigenous tribespeople, ethnic minorities, and other oppressed groups—a point that is of crucial importance if we are to develop a *general* ecocentric emancipatory theory. Here, ecofeminist theorists need to be wary of the problem of over-identifying with, and hence accepting uncritically, the perspective of women. Such an over-identification can sometimes *inhibit* the general emancipatory process by offering an analysis that can (i) deny the extent to which many women may be complicit in the domination of nature; (ii) overlook the various ways in which men have been oppressed by limiting "masculine stereotypes"; and (iii) be blind to other social dynamics, institutions, and prejudices that do not bear on the question of gender. Moreover, privileging—rather than simply rendering visible and *critically* incorporating—the special insights of women can sometimes lead to a lopsided and reductionist analysis of social and ecological problems.

For example, an uncritical identification with the female stereotype can sometimes lead to a simple reversal of the human characteristics that are considered to be valuable, that is, a replacement of the hyperrational, impersonal, and abstract "male" standard of human excellence with an excessively particular, personal, and emotional "female" standard. Indeed, many ecofeminists, drawing on theories of gender and self-development (such as object relations theory), have argued that the "feminine" sense of self is preferable to the "masculine" sense of self on the ground that it gives rise to a personal, reciprocal, emotional, and contextualized "caring ethic" as opposed to an abstract, rights-based ideal of justice.[83] While we certainly need to discredit the masculine stereotype we also need to be wary of certain aspects of the feminine stereotype. Indeed, if the speculative theory of object relations tells us anything, it is that the prevailing stereotypical male and female senses of self are both deficient—that the former is excessively delineated whereas the latter is *not delineated enough*.[84] This would seem to undermine the claim that a new environmental ethic ought to speak in the "different voice" of women (to adopt Carol Gilligan's phrase, although this is not her claim), suggesting instead that both the masculine and feminine stereotypes need to be transcended (at least in part and not necessarily to the same degree), not only by shared parenting but also by other kinds of social and cultural change.[85]

A further tendency toward reductionism is illustrated by the argument maintained, or implied, by some ecofeminists that it is patriarchy that lies at the root of the domination of women *and* nature. This argument suggests that the principal focus of an emancipatory *ecological* praxis must be patriarchy rather than anthropocentrism (indeed, this charge has been levelled against deep ecology).[86] However, it is one thing to note parallels in the logic or symbolic structure of different kinds of domination (surely this is enough to explain the strong resonance in the egalitarian orientations of the radical feminist and ecology movements) and another thing to argue that the kinds of domination that radical feminists and radical ecologists are addressing stem from the *one* source. To maintain this causal explanation, it must be shown that patriarchy not only *predated* but also gave rise to anthropocentrism—in other words, that there is a necessary connection between the two phenomena. How, then, do we explain the existence of patriarchy in traditional societies that have lived in harmony with the natural world?[87] How do we explain Engels' vision of "scientific socialism," according to which the possibility of egalitarian social/sexual relations is premised on the instrumental manipulation and domination of the nonhuman world?[88] Clearly, patriarchy and the domination of nonhuman nature can each be the product of quite different conceptual and historical developments. It follows that the emancipation of women need not necessarily lead to the emancipation of the nonhuman world and vice versa.

The above criticisms are not intended to deny that patriarchy and anthropocentrism can be mutually reinforcing when they do occur together. In this respect, both women and the nonhuman world can indeed be seen to have a mutual "interest" in emancipation from the status of Otherness. Moreover, at the symbolic and conceptual levels, both women and nonhuman nature have been associated and downgraded in the God-Man-Woman-Nature hierarchy of being—a conceptual schema that has served to *legitimize* the greater social status and power held by men vis-à-vis women and nonhuman nature. However, as Fox has argued, a variety of human/nonhuman distinctions have served as the fundamental legitimating ideology not only for patriarchy but also for other kinds of human oppression. For example, the fundamental legitimating ideology of racism and imperialism has been that whites and Westerners are seen to possess—or possess to a greater extent than their counterparts (i.e., blacks, or non-Westerners)—certain qualities that are deemed to be of the essence of humanness (e.g., rationality, civilization, or being more favored by God). In other words, not only men but also whites and Westerners (both men and women) have sought to legitimate their superior social position by claiming that they are somehow "more fully human" than, and hence morally superior to, women, blacks, and non-Westerners.[89] Of course, as Val Plumwood has pointed out, the success of this kind of legitimation in relation to patriarchy depends on a general acceptance of a concept of the human that is set apart from the rest of the animal world on a hierarchy of being.[90] Replacing such a hierarchical mode of perceiving the world with an ecological sense of self that affirms others (both humans and nonhumans) in a state of reciprocal interdependence serves to undermine the conceptual apparatus that has legitimated not only patriarchy and other forms of human oppression but also anthropocentrism. In other words, patriarchy may be seen as not the root of the ecological crisis but rather a subset of a more general problem of philosophical dualism that has pervaded Western thought (e.g., mind/body, reason/emotion, human/nonhuman) from the time of the classical Greek philosophers.[91]

Such a recognition has led many ecocentric theorists (including transpersonal ecologists and many ecofeminists) to argue that we need to transcend masculine and feminine stereotypes and cultivate a new kind of *person* that possesses "the *human* characteristics of gentleness and caring" [my emphasis].[92] This does not mean that the new ecological person must be thoroughly androgynous or gender neutral (there is no reason why the differences between men and women should not be celebrated), *only that a person's sex is not considered to have an important bearing on the human qualities that are needed to heal the rift between humans and the rest of nature.* As Don E. Marietta has put it, "We are talking about people who cultivate the best qualities of human beings, regardless of the traditional assignment of those to one sex. These qualities of character and behaviour indicate, I believe, the values supported by feminism."[93] Similarly, Val Plumwood has argued that what we now need is "an account of the human ideal for both sexes, which accepts the undesirability of the domination of nature associated with masculinity."[94] Such an ideal must flow from a critique of *both* masculinity and femininity and be linked to a "systematic transcendence of the wider set of dualisms" (e.g., mind/body, reason/emotion, public/private).[95]

To recapitulate, then, ecofeminists have drawn attention to the conceptual parallels, symbolic resonances, and areas of practical overlap in both the critical and constructive tasks of the radical wings of the ecology movement and the women's movement. However, in those areas where ecofeminism diverges from a deep/transpersonal ecology perspective (i.e., in emphasizing purely personally based forms of identification; in sometimes *uncritically* privileging the experience of women; and in sometimes overstating the links between patriarchy and the domination of nature) it is vulnerable to criticism. Of course, not all ecofeminists make these claims and, of those who have, not all have pursued them in the same way or to the same end. Indeed, given that there are many different kinds of feminism (e.g., liberal, Marxist, social-ist, radical, existentialist, psychoanalytical, and postmodern), it is hardly sur-prising to find that there is more than one kind of ecofeminism.

However, when we turn to those particular expressions of ecofeminism that are *not* vulnerable to the above criticisms, we find an ecophilosophical orientation that is almost indistinguishable from that of transpersonal ecolo-gy. That is, we find a similar ecophilosophical orientation of inclusiveness, albeit from an "inside-out" rather than an "outside-in" route. The qualifica-tion "almost" remains important, however, when it comes to deciding on an ecophilosophical framework and label for a *general* emancipatory theory. Such a framework must be able to critically incorporate the special experi-ences and perspectives of all oppressed social groups, not just women. Now an ecofeminist perspective is quite capable of doing this—indeed, Karen Warren has defended an inclusive and pluralistic version of ecofeminism that is concerned to end *all* forms of oppression (and which provides a general ecocentric emancipatory theory).[96] Yet some ecofeminists might want to argue that ecofeminism *should* remain a specifically feminist project in the sense of providing a special voice to women in view of the pervasiveness of patriarchal culture and its links with the domination of nature. (In any event, this is what is suggested by the label "eco*feminist*.") This is, I believe, a valid case, although it would mean that ecofeminism would then become a major and essential tributary of a general ecocentric emancipatory framework rather than serve as *the* general emancipatory framework. Moreover, a case might be made that a general ecocentric emancipatory theory must be one that in both purpose *and* name does not privilege the concerns of any particu-lar human emancipatory movement.

Whatever label is ultimately adopted, however, a general ecocentric emancipatory theory must accommodate all human emancipatory struggles within a broader, ecological framework. That is, it must be able to provide the context for establishing the outer ecological limits within which the dif-ferent needs of human emancipatory movements can be addressed and har-monized in order to ensure that the interests of the nonhuman world are not continually sacrificed in the name of human emancipation. The emancipatory concerns of new social movements—including the women's movement—may thus be seen *as nesting within* such an ecocentric framework. Transper-sonal ecology provides one such general theoretical articulation of ecocen-trism. As we have seen, autopoietic intrinsic value theory provides a complementary general theory of ecocentrism. Ecofeminism can provide, at a minimum, an essential component of a general ecocentric theory by pointing to the links between the domination of women and the domination of nature, or, alternatively, it can be formulated in such a way as to provide a general theory of ecocentrism. However, all of these approaches provide only the *ecophilosophical* underpinnings of ecocentrism. It remains to explore how this inclusive ecophilosophical framework might be fleshed out in a political and economic direction.

Part II

*An Ecocentric Analysis
of Green Political Thought*

CHAPTER 4

The Ecocentric Challenge to Marxism

INTRODUCTION

The pressing nature of the environmental crisis and the growing political prominence of the environmental movement in the last three decades have prompted many Marxist theorists to turn their attention to the relationship between environmental degradation, capitalism, and social justice. More recently, Marxist scholars have had further cause to embark upon a critical examination of Marxist theory and practice as news of ecological devastation in Eastern Europe continues to mount in the wake of the dismantling of the Iron Curtain.

The Marxist entry into the ecopolitical debate marks the beginning of an interesting new phase in the development of Marxist thought, the outcome of which may well determine the extent to which it is able to exert a continuing influence on political movements in the closing decade of this century and beyond. I say "new phase" here because environmental degradation has not been a traditional concern of Marxism. Indeed, it has generally been considered a mere epiphenomenon of capitalism rather than important in its own right—something that, in any event, will be brought under rational social control in a socialist society. This has been reflected in the widespread tendency among Western Marxists, especially in the "early" days of modern environmental concern (i.e., the late 1960s and early 1970s), to dismiss environmentalism as an elitist preoccupation of the middle class who can "afford" to worry about such matters.[1] In particular, calls by radical environmentalists and Green parties for the curbing of economic growth have been met with considerable suspicion for raising what were seen as repugnant neo-Malthusian arguments that were blind to the inequitable social consequences that were presumed to flow from a scaling down of production.

Although the Marxist critique of the middle class character of environmentalism has formed an important part of ecopolitical debates in the West, I have addressed these charges elsewhere and do not intend to evaluate them again here.[2] My main purpose in referring to these debates is simply to draw attention to the essentially reactive and defensive character and tone of the initial entry into the environmental debates by Marxists. To many Marxists and working-class sympathizers, the environmental movement was initially perceived more as a threat to the political and theoretical terrain staked out by the Left—as a backsliding toward conservatism—than a harbinger of novel and progressive political ideas.

Although some Marxists have adopted a purely defensive posture, a growing number of Marxist sympathizers have undertaken the more constructive task of developing a specifically Marxist response to the environmental crisis. The upshot is that Marxism now stands not simply as a source of external criticism against the Green movement; there are many who argue, for a variety of reasons, that the Green movement ought to embrace a Marxist perspective as an alternative to what is seen as the utopianism, idealism, and "voluntarism" of much Green theorizing. What is needed, these critics argue, is a more materialist approach that is cognizant of the relationship between class and the inequitable impact of environmental degradation and is prepared to directly challenge institutionalized power relations in society. Indeed, for some Green theorists, particularly in Europe, Western Marxism has served as the theoretical starting point for their analysis of the ecological crisis.[3]

The injection of various Marxist perspectives into ecopolitical debates has helped to facilitate a fruitful political dialogue between the labor and environmental movements, which has, in turn, served to challenge and widen both perspectives. Moreover, the increasing public prominence of ecological issues alongside new developments in the growing field of environmental philosophy, both of which underscore the importance of preserving wilderness and ecological diversity, has thrown down significant new challenges to Marxist theory. Indeed, we shall see that it has been the ecocentric arguments for wilderness preservation that Marxist scholars have found to be the furthest removed from their traditional concerns and consequently the hardest to assimilate into their theoretical framework.

However, it must also be borne in mind that Marx formed his ideas at a time when he could not have been expected to anticipate the extent of global ecological degradation that we now face. It is indeed testimony to the stature of Marx and the hold of his ideas that scholars should still be seeking illumination and direction from his writings in respect of problems that he regarded in his day as epiphenomenal. Of course, most ecocentric Green theorists consider this to be a good reason in itself to abandon Marxism as being ill equipped to tackle contemporary environmental problems. According to these theorists, it is not necessary to adopt a Marxist perspective (revised or otherwise) in order to acknowledge, say, the importance of an equitable sharing of the costs of environmental reforms or the many ways in which the profit motive and the dynamics of capital accumulation have contributed to our current environmental ills. From the point of view of these theorists, the intellectual legacy of Marxism is seen to be fundamentally incompatible with an ecocentric perspective.[4]

Recent efforts to redress Marxism's traditional neglect of environmental issues have been conducted on two major levels: first, a rereading of the writings of Marx and Engels in order to discern their attitude toward nature and technology and to find out how and to what extent they addressed the environmental problems of their day;[5] and, second, an inquiry into the question as to whether Marxism can be developed in such a way as to address constructively the environmental crisis in terms that are relevant to the conditions of the late twentieth century.[6] Significantly, some of the Marxist scholars who have embarked on this new area of inquiry have become increasingly critical of certain aspects of Marxist theory. In some cases this has occurred to the point where these theorists have either taken on a new ecosocialist label (André Gorz) or become recognized as post-Marxists or ecoanarchists (Rudolf Bahro).[7] These more radical revisions and departures will, however, be explored in chapters 6 and 7. My concern in this chapter is with the theoretical contributions of those scholars who have found sufficient illumination in the original writings of Marx to enable them to develop a Marxist explanation of, and response to, the ecological crisis. It should be pointed out that the discussion of orthodox Marxism and its ecological off-shoot (orthodox eco-Marxism) is presented here merely as a foil or point of departure for, rather than as an example of, Green political theory.

THE THEORETICAL ROOTS

Although Marx was only marginally concerned with environmental degradation in his day and although he did not present a systematic theory of humanity's relationship to nature, there exist numerous passages in his wide-ranging *oeuvre* that enable his position on both of these matters to be discerned.[8] The overriding sense in which Marx characterized nature was as a *medium for human labor,* as the means by which the power of the human laborer could be revealed, or as Alfred Schmidt has summarized it, "as the means and the material of [our]...self-realization in history."[9] The nonhuman world, which Marx often referred to as "external nature," was first and foremost "the primary source of all instruments and objects of labor" and was variously referred to as a "laboratory," "the original tool house," or the "original larder."[10] Like Locke before him, Marx accepted the view that the mixing of human labor with external nature was an act of appropriation, that the product belonged to the laborer, indeed, revealed the *power* of the laborer since labor and its extension—technology—brought about what nature could not accomplish alone. Although Marx saw humans as a part of (rather than separate from) nature, human labor was nonetheless seen as playing a pivotal and determinative role in nature's unfolding. Drawing on Hegel, Marx portrayed the labor process as a "metabolism" between humans and external nature with humans as Subject confronting external nature as Object in a dialectical movement that led to the transformation of both humans and the nonhuman world. Humanity and external nature were characterized as two indivisible "moments" in nature's self development.[11]

In tracing the development of Marx's theoretical perspective on nature, it is possible to find both continuities and discontinuities between the writings of the young (i.e., pre-1845) Marx and the mature and more economically preoccupied Marx of *Capital*. In the so-called Paris Manuscripts (i.e., *The Economic and Philosophical Manuscripts of 1844*), Marx referred to the labor process as effecting the progressive "humanization" of nature and "naturalization" of humanity. Nature was described as "the inorganic body" of humanity that had been increasingly assimilated, through work, into an "organic" part of humanity.[12] In his essay on estranged labor, Marx described humanity's transformation of the external world through labor as the means by which humanity realized its "species being" (or human essence)—a notion that Marx took from Ludwig Feuerbach.[13] Marx argued that the relationship between humanity and external nature is a transformative one whereby both humans (and their needs) and external nature are changed as the labor process expands humanity's productive powers.

According to Marx, whereas (nonhuman) animals produce one-sidedly, that is, only for their immediate needs, humans produce universally, that is, even when free from physical need "and only truly [produce] in freedom thereof."[14] Marx argued that alienated labor degrades to a mere means of physical existence what might otherwise be spontaneous, free human activity. As a result, humans become estranged from their labor and its product, which confronts them as an alien Other at the behest of the masters of labor—the owners and controllers of capital. Marx saw this as also leading to the estrangement of humans from themselves, from their fellow humans, from nature, and from their species being.[15] Marx's solution to the problem of alienated labor was the revolutionary transformation of the institutions of wage labor and private property via the expropriation by the proletariat of the capitalist means of production—a transformation that Marx envisaged as paving the way for a fully *social* mastery of nature.

Marx's treatment of humans as *homo faber* is a central feature of the antagonistic dialectic between humanity and nature set out in these early writings. Labor and its extension—technology—were seen not only as a means to survival but also as *the* road to human self-realization. History was seen as the progressive humanization of nature and naturalization of humanity resulting in an ever greater equivalence between humanity and nature (i.e., where nature increasingly appeared as made rather than given, domesticated rather than wild). Communism was to be that stage where individuals would live in a classless society and be free to engage in self-determining activity, since they would no longer be dominated by the functional economic imperatives of capitalism, the commands of a dominant class, *or external nature.* It would also be that stage where the human/external nature dialectic would be reconciled via the *complete* "humanization" of nature. Through technological innovation and automation, and the subordination of economic and natural processes, humans would thus recover *time* in which to enjoy freedom beyond the dull compulsion of labor.[16] In short, the young Marx believed that the "realm of necessity" would give way completely to the "realm of freedom" in a communist society.

While the notion of humans as *homo faber* remained a central theme in the writings of the mature Marx, he came to the view "that the struggle of man with nature could be transformed but not abolished."[17] That is, the complete "reconciliation" with, or "humanization" of, nature was no longer considered possible, because although labor could be reduced to a minimum, Marx took the view that it could never be totally dispensed with. Although more and more areas of nature would come under human control through technological development, the *antagonistic* dialectic between humanity and nature would never be entirely resolved.[18]

Another significant change in emphasis in Marx's writings concerned the theme of alienation. In particular, from the time of Marx's first systematic presentation of historical materialism (namely, the critique, jointly authored with Engels, of German idealism and the Young Hegelians in *The German Ideology* [written in 1845–46]) he no longer emphasized concepts such as "alienation" or the "realization of the human essence."[19] Instead of explaining historical change in accordance with a philosophical concept of "species being," Marx became increasingly preoccupied with analyzing the dynamics of the capitalist mode of production—an analysis that he referred to as an "objective" and "scientific" study of the unfolding of history. By the time of the publication of *Capital,* Marx had consolidated the now familiar distinction (which was frequently conflated in his earlier work) between the *forces* and *relations* of production. The forces (or means) of production were understood as the technological means with which humans control external nature in order to satisfy their needs, including tangible means such as machines as well as more intangible means such as scientific knowledge, skills, and organizational layout. The relations of production referred to that ensemble of property, class, and legal arrangements that regulated the control and ownership of the production process and the distribution of its fruits. In capitalist societies, these relations of production were seen as relations of exploitation where unpaid surplus-labor was siphoned off to the capitalist class. Together, the forces and relations of production formed an articulated unity—the *mode* of production—that corresponded respectively to the technical and social determinants of humanity's interaction with external nature.

According to Marx, the exploitation embodied in the capitalist relations of production could be overcome only by changing these relations. However, Marx's contempt for the relations of production did not extend to the rapidly expanding forces of production of his time. Quite the contrary, he extolled the new techniques of industrial society as the harbinger of freedom in creating the material and social preconditions for a socialist society. Total automation (considered a definite advance over handicrafts) would, under revolutionized *relations* of production, free the laborer to enter a qualitatively different relationship to the production process as overseer and regulator. Marx believed this process would lead to the development of more rounded individuals rather than specialized ones. Marx also welcomed what he saw as the civilizing influence of technology and rejected nature romanticism and primitive cultures alike as "childish," "backward," and "reactionary" in opposing, or otherwise showing no inclination toward, technical progress.[20] Marx fully endorsed the "civilizing" and technical accomplishments of the capitalist forces of production and had thoroughly absorbed the Victorian faith in scientific and technological progress as the means by which humans could outsmart and conquer nature. Indeed, Marx welcomed the challenge thrown down to humans by a "stingy" nature:

> This mode [i.e., capitalism] is based on the dominion of man over Nature. Where nature is too lavish, she "keeps him in hand, like a child in leading-strings." She does not impose upon him any necessity to develop himself.[21]

As Balbus has observed, Marx saw the development of science as the means by which humanity would seek to "discover nature's 'independent laws' not in order to respect, but rather in order to undermine, its independence from our existence."[22] Conquering nature in this way was welcomed as the means of human self-aggrandizement, for Marx considered that, once outsmarted by "man," external nature "becomes one of the *organs* of his activity, one that he annexes to his own bodily organs, adding stature to himself in spite of the Bible."[23]

Although Marx changed his view concerning the extent to which humanity would be able to master necessity, he consistently saw human freedom as inversely related to humanity's dependence on nature. Moreover, he argued that human freedom properly began only when mundane, necessary labor ceased:

> Freedom...can only consist in socialised man, the associated producers, rationally regulating their interchange with Nature, bringing it under their common control, instead of being ruled by it as by the blind forces of Nature; and achieving this with the least expenditure of energy and under conditions most favorable to, and worthy of,

their human nature. But it nonetheless still remains a realm of necessity. Beyond it begins that development of human energy which is an end in itself, the true realm of freedom, which, however, can blossom forth only with this realm of necessity at its basis.[24]

Engels also endorsed this understanding of the free communist human being. He believed that once the capitalist relations of production had been overthrown, humans would *ascend* from "the kingdom of necessity to the kingdom of freedom."[25] In the *Dialectics of Nature,* Engels considered the essential distinction between humans and nonhumans to rest on humanity's ability not simply to use and change nature (something all animals do) but also to *master* it by making it serve humans ends.[26]

Yet Engels also shrewdly noted that we ought not "flatter ourselves overmuch on account of our human victories over nature. For each such victory takes its revenge on us."[27] In tracing the development of the forces of production, Engels showed a keen awareness of the many unintended ecological dislocations brought about by the laboring activities of humans in both industrial and preindustrial times. Moreover, he observed that

at every step we are reminded that we by no means rule over nature like a conqueror over a foreign people, like someone standing outside nature—but that we, with flesh, blood and brain, belong to nature, and exist in its midst, and that all our mastery of it consists in the fact that we have the advantage over all other creatures of being able to learn its laws and apply them correctly.[28]

Similarly, there are frequent passages in *Capital* where Marx also observed how the dynamics of capital accumulation had led to the exploitation of the laborer and soil alike.[29] Yet both Marx and Engels continued to welcome the powerful forces of production unleashed by capitalism for creating the "material conditions for a higher synthesis in the future" where further advances in the natural sciences would enable humans to predict and control the more remote consequences of their increasing incursions into nature.[30] In Marx's lexicon, emancipation meant being freed from both social and "natural" oppression (the latter made possible by the mastery of nature's laws, which included the control and containment of the unwanted ecological "side effects" of human productive activity).

As we shall see, Marx's juxtaposition of freedom and necessity, the former corresponding to the mastery of social and natural constraints and the latter corresponding to subservience to social and natural constraints, has remained an enduring theme in ecosocialist thought, particularly in the work of Herbert Marcuse and André Gorz.

The recent efforts to develop a Marxist solution to the environmental crisis may be divided into two streams in accordance with the convenient distinction between "humanist" and "orthodox" Marxism (which loosely maps onto the work of the "young" and "mature" Marx respectively). The humanist eco-Marxists have sought to develop a more ecologically sensitive Marxist response to the environmental crisis that seeks to harmonize relations between the human and nonhuman realms. Orthodox eco-Marxists, on the other hand, make no apologies for being anthropocentric and are critical of humanist eco-Marxists for being idealist, voluntarist, and decidedly "un-Marxist." From an ecocentric perspective, however, it will be shown that *both* streams of eco-Marxism uncritically accept Marx's view of history and his particular notion of humanity as *homo faber* and thereby perpetuate an instrumentalist and anthropocentric orientation toward the nonhuman world.

ORTHODOX ECO-MARXISM

Orthodox eco-Marxists have strayed very little from the basic position of the "mature" Marx set out in the preceding section.[31] That is, environmental problems, like social problems, are traced directly to the exploitative dynamics of capitalism. The solution to these problems is seen to require the revolutionary transformation of the *relations* of production combined with the development of a better theoretical understanding of nature and further advances in technology so that a complete social mastery of nature can be attained for the benefit of all, rather than just the privileged capitalist class. This orthodox eco-Marxist interpretation thus retains Marx's view of history as a *progressive* dialectical struggle from the primitive to the advanced, resulting in the increasing domestication of the nonhuman world through the activity of labor and its extension, technology. In view of this, as Tolman explains,

> it should be clear why Marxists should continue to support the development of science and technology, and why they should assert the ultimate unity of science, technology, the mastery of nature, and humanism. Taking human history as a dialectical whole, these can all be seen as essential components of human nature itself. If understood in this dialectical sense, the Marxist gladly accepts the charge of "homocentrism."[32]

To orthodox eco-Marxists, the setting aside of areas of wilderness for the protection of endangered species and the preservation of biotic diversity in general can be justified only if it can be shown to be of some *instrumental* value to humans, by providing, say, a place of recreation or a store of potential raw materials for humanity's future productive labor. According to orthodox eco-Marxists, it simply makes no sense to say that the nonhuman world *ought* to be valued and protected for its own sake. For example, Howard Parsons, in his exhaustive review (and endorsement) of Marx and Engels' position on nature and technology, has trouble in grasping what the case against anthropocentrism is all about: "It is hard to know," he confesses, "what could be meant by nature 'in itself,' either in a Kantian sense or in the sense of a discrete reality entirely independent of our cognition and action."[33] On the basis of this epistemological point—that nature cannot exist independently from our values and actions—Parsons concludes that humans cannot *value* the nonhuman world for its own sake.

Yet Parsons' answer to the critique of anthropocentrism misses entirely the *normative* point of the ecocentric critique. First, he commits what Fox refers to as "the anthropocentric fallacy" in that he conflates the trivial and tautological sense of the term anthropocentric (i.e., that all our views are, necesssarily, human views) with the substantive and informative sense of the term anthropocentric (the unwarranted, differential treatment of other beings on the basis that they do not belong to our *own* species). Second, and in any event, the ecocentric argument is not that we should value the nonhuman world because it exists independently from human values and actions. Ecocentric theorists would be the first to agree that there are no absolute divides in nature, that we are connected, in varying ways, with the nonhuman world and vice versa. However, despite these interconnections, the model of internal relations that informs ecocentrism also recognizes the *relative autonomy* of all entities. On the basis of this recognition, ecocentric theorists are concerned to cultivate a prima facie orientation of nonfavoritism that allows both human and nonhuman entities to unfold in their own ways. And it is precisely because we are part of an interconnected, larger whole that ecocentric theorists argue that we should exercise our own freedom with care and compassion.

However, it is clear from Parsons' discussion that even if he properly grasped the ecocentric argument he would still reject its normative and practical claims. For example, Parsons rejects as "unrealistic" the radical ecological argument that we should simplify human needs, reduce human population and consumption, respect nature, and lead a more agrarian lifestyle. (All but the last of these points more or less reflect the kinds of changes defended by many ecocentric theorists; I argue in chapter 7, however, that it is neither necessary nor desirable that everyone live in decentralized, rural settlements and there is a strong case for urban settlements in an ecocentric society.) According to Parsons, human survival and well-being depend on a "knowledge and control of nature's substances and processes."[34] Yet the kind of knowledge Parsons and other orthodox eco-Marxists have in mind is not the kind that sees nature as a pattern or set of interrelationships to respect and follow or a design from which to draw guidance and inspiration. Rather, it is knowledge that will enable humans to overcome and redirect any resistance to their struggle for total mastery of nature. As Parsons explains,

> Marxism rejects this [ecocentric] unrealism, and in its view of the man-nature relation, it is inclined to emphasize man rather than the

plants and the animals. If the assumption of the criticism is that Marxism has this emphasis, the assumption is correct. And like many modern humanisms, Marxism has sometimes overemphasized man's place in nature.[35]

Clearly, orthodox eco-Marxists regard ecocentrism as putting an unnecessary restraint on human development, which they regard as dependent upon an expanding science and technology that will increase our ability to control and manipulate the "secrets of nature." This, of course, is entirely consistent with Marx's exclusive preoccupation with human betterment. Marx showed no interest in natural history, and he did not address the cause of nonhuman suffering. Indeed, Parsons defends Marx and Engels' lack of interest in the emerging "humane societies" for the prevention of cruelty to animals in the eighteenth and nineteenth centuries, arguing that the concern for the welfare of nonhuman animals was "a displacement of human concern" that was restricted by the privileged class position of its advocates.[36] In any event, Parsons suggests that in the long run nonhuman animals, like human animals, might also be liberated by technology:

> Presumably when animals are displaced entirely by machines as instruments of production, and when food is synthesized chemically, animals will enjoy a freedom not enjoyed since their domestication for food and labor in Neolithic times, and man's attitude toward them will likewise change with man's new freedom.[37]

However, far from being displaced by machines, many domestic animals (most notably cattle and hens) have been effectively *turned into* machines as a result of the application of "advanced" production techniques in agriculture. Moreover, ecological reality suggests that unless human population growth and the loss of genetic diversity and wild habitat is drastically curbed, there will be a very narrow range of nonhuman animals left to enjoy the distant "freedom" that Parsons believes will be wrought by advanced technology!

As we have seen, to the extent that environmental problems were acknowledged by Marx and Engels, they were attributed to the capitalist relations of production, not the forces of production. Orthodox eco-Marxists have fully endorsed this analysis of the problem: the capitalist classes, while initially facilitating the development of the productive forces, are ultimately seen as acting as a fetter to their full development by standing in the way of a complete social appropriation of the power of nature and the control of any unwanted side effects. According to Parsons:

> an economic system [such as the capitalist one] that breaches the laws of nature by which wealth is produced will bring on inevitable reactions: impairment of nature's "metabolism" of ecological

cycles, depletion of nature's resources, impoverishment of human society, *and a relapse of nature into the slumber of undevelopment* [my emphasis].[38]

The Marxist explanation for ecological degradation lies in the fact that capital works only for the benefit of the owners and controllers of capital rather than for the benefit of the complete society of producers. (It is assumed that the complete society of producers would act as a collective and would therefore be concerned to protect public environmental goods such as air, water, and soil). It is thus the dynamic of *private* capital accumulation that has given rise to resource depletion, pollution, untrammelled urbanization, and the occupational and residential hazards suffered by workers and their families.[39] Yet as John Clark observes, although Parsons tries to establish Marx's ecological credentials by arguing that Marx recognized an "essential incompatibility" between capitalism and "the system of nature," Parsons in fact misses Marx's anti-ecological point:

> Marx's point is not that this expansionism is in conflict with nature, but rather that capital's quest for surplus value contradicts and *limits* this development in some ways, to the detriment of humanity. In contrast, an ecological critique would question this very expansionism as being in contradiction with "the system of nature" [my emphasis].[40]

Orthodox eco-Marxists simply seek to replace the private and socially inequitable mastery of nature under capitalism with the public and socially equitable mastery of nature under communism. Ecological degradation under capitalism is seen by orthodox eco-Marxists as a measure of its *inefficiency*, of its failure to utilize resources wisely. As such, orthodox eco-Marxists are predominantly resource conservationists; they are at home with Gifford Pinchot (with his "wise use" of natural resources argument), but are fundamentally at odds with John Muir's vision of large tracts of wilderness being protected in their "state of natural grace."[41] As we have seen, to the extent that orthodox eco-Marxists would be prepared to defend ecosystem preservationism, it would be on purely human-centered, instrumental grounds. Of course, Marx and Engels were also early pioneers of what I have called the human welfare ecology stream of environmentalism. Indeed, Engels's classic critique of the working and living conditions of the Victorian working class is a major milestone in the development of this stream of environmentalism.[42] However, while Marx and Engels's critique challenged capitalism and exposed the misery of the working class, it did not challenge the hegemony of instrumental reason. (This challenge has, however, been taken up by some Western Marxists, as we shall see in the following chapter.)

From an ecocentric perspective, the "true freedom" promised by scientific socialism is ultimately illusory. As Bookchin puts it, "at its best, Marx's work is an inherent self-deception that inadvertently absorbs the most questionable tenets of Enlightenment thought into its very sensibility."[43] Moreover, ecofeminists and social ecologists have drawn attention to the "masculine" character of the mastery sought by the mature Marx, who rejected as regressive the idolization of nature as a "nurturing mother." According to Marx, modern "man" must sever his umbilical cord with nature and become a self-determining being in order to achieve his "manhood." As John Clark has observed,

> Marx's Promethean and Oedipal "man" is a being who is not at home in nature, who does not see the Earth as the "household" of ecology. He is an indominable spirit who must subjugate nature in his quest for self-realization....
> For such a being, the forces of nature, whether in the form of his own unmastered internal nature or the menacing powers of external nature, must be subdued.[44]

Finally, orthodox eco-Marxists continue to place faith in the working class as the agents of revolutionary change, both social and environmental. According to Parsons, the proletariat still remain the class best situated for "assuming a position of ultimate power and responsibility over the whole transformed system"—including control of the unwanted side-effects of the manipulation of nature.[45] Post-Marxist Green theorists, in contrast, have challenged what they call the "productivist ideology" and inherent conservatism of the Western labor movement and have pointed instead to the radical potential of new social movements, particularly those concerned with ecology, feminism, and Third World solidarity.[46]

None of the above criticisms are intended to dismiss the eco-Marxist critique of capitalism. Indeed, this critique still has relevance today in terms of highlighting the ways in which capitalism can exploit laborer and land alike, particularly in Third World countries where ecological considerations are displaced almost entirely in the drive to develop massive power schemes and large scale export industries (often merely to service the large debts owed to First World countries).[47] From an ecocentric perspective, however, the eco-Marxist critique simply does not go far enough: it is fundamentally limited by its anthropocentrism, its focus on the relations of production at the expense of the forces of production, and its uncritical acceptance of industrial technology and instrumental reason.

Not surprisingly, the shortcomings of orthodox Marxism have attracted the critical attention of a number of ecologically concerned Western Marxists of a more "humanist" persuasion who are critical of "scientific socialism" yet still attracted to Marxism as a philosophy and form of critique.

Unlike orthodox eco-Marxists, humanist eco-Marxists argue that it is necessary to reassess Marx's technological optimism and nineteenth century belief in material progress. According to André Gorz, the collective appropriation by the proletariat of the capitalist forces of production would not solve the ecological crisis: it would simply mean that the proletariat would take over the machinery of domination.[48] Gorz has argued that the development of the productive forces, hitherto welcomed by most Marxists, must be reexamined on the grounds that they now threaten society's ecological support system."[49] Similarly, Enzensberger has argued that the ecological crisis can be dealt with in Marxist terms if we remember that capitalism is not just a property relation but also a *mode of production* in which the forces and relations of production are inextricably linked; this capitalist mode of production is something that the socialist countries, which he considers to be "still in transition," have yet to abandon.[50]

In asking how we might resolve the ecological contradictions of capitalism and what kind of human being will inhabit the society that lies beyond the realm of domination, humanist eco-Marxists have sought inspiration from the philosophical writings of the young Marx.[51] Indeed, John Ely has shown that certain aspects of the young Marx's utopianism concerning the reconciliation of humanity and nonhuman nature have been taken up directly in the West German Greens' first economic program, although the usage of these concepts by the Greens is selective rather than systematic.[52]

The most ecologically sensitive case for a return to the ideas of the young Marx is that provided by Donald Lee in his essay entitled "On the Marxian View of the Relationship between Man and Nature."[53] However, although Lee's particular vision of humanist eco-Marxism is, for the most part, no longer vulnerable to the criticisms that I have levelled against orthodox eco-Marxism, a case can be made that it is also no longer *Marxist*. That is, Lee has developed a particular version of humanist eco-Marxism that downplays, and in some cases ignores, key distinctions and themes in the writings of the young Marx. Yet it is precisely these particular distinctions and themes (which, as we shall see, are endorsed by *other* versions of humanist eco-Marxism) that make humanist eco-Marxism incompatible with an ecocentric perspective. The most important of these is Marx's distinction between "freedom" and "necessity." As we shall see, this key distinction serves to make the domination of the nonhuman world a *requirement* of human self-realization.

The basic goal of Lee's approach is to overcome alienation in a very broad sense, with the ecology crisis taken as evidence of our *alienation from nature* and as one more obstacle in the path to human emancipation. According to Lee's reading of Marx's early writings on alienation, there is no human/nature (or Subject/Object) dichotomy in Marx's thinking, as is often claimed. Rather, Lee argues that Marx was concerned to overcome alienation between humans and themselves and their work *and between humans and external nature.* Under capitalism, both the worker *and* nonhuman nature are considered by the capitalist class as instruments to be exploited for private profit. According to Lee, "Capitalism was a necessary stage in man's development of the mastery of nature: but a further development is now necessary, namely, the overcoming of the dichotomy between man as subject over and against nature as object."[54]

In Lee's view, Marx's ideas were not anthropocentric because he conceived of nature as humanity's *inorganic body*: "This recognition of nature as our *body* will constitute the overcoming of the alienation of ourselves from nature, manifested in subject-object dualism."[55] On the basis of this insight, Lee has sought to outline an ecologically benign form of socialist stewardship of nature that will emancipate humans and nonhumans alike from the tyranny of capitalism so that humans can become the caretakers of their own "body." This socialist notion of nonhuman nature as *our* inorganic body, toward which we have a responsibility of care, is juxtaposed to the capitalist conception of nature as an alien "other" to be exploited for private profit. According to Lee, an ecological ethic must become part of the Marxist program of liberation; socialism must be developed to what Lee sees as its logical end, that is, beyond homocentrism (i.e., anthropocentrism).[56]

Lee argues that the present dichotomy between humanity and nature may be overcome only through the overthrow of the wasteful capitalist system (read "mode of production") and its replacement by "a rational, humane, environmentally unalienated social order."[57] The major features of this new order would be socially useful production, the reduction of labor time, maximum creative leisure, wise use of resources, rational population control, and solidarity between all living things, not just humans. Lee's post-scarcity utopia is thus one in which everyone will be able to realize what Marx referred to as our "species being"—a situation of genuine freedom from need—whereby we can create according to the "laws of beauty" and become responsible stewards of the ecosystem. According to Lee, this form of socialist stewardship is based on a sense of enlightened self-interest because:

> man is the *universal* being who can understand what is good for each species intrinsically, and thus, just as socialist man transcends the selfish greed of the capitalist and acts for the good of all men (which is ultimately his own good) so must ecologically aware socialist man transcend the selfish greed of homocentrism and act for the good of the whole ecosystem (which ultimately is his own good).[58]

Lee's socialist post-scarcity utopia represents a considerable departure from the orthodox eco-Marxists' Promethean orientation toward nonhuman nature. That is, Lee's socialist society does not seek to dominate nonhuman nature as an alien "other" through the development of technology. Rather, it is to be a society of self-determining individuals who realize themselves through free, conscious activity and who recognize nonhuman nature as but an extension of themselves, as part of their inorganic body.

However, what is *not* apparent in Lee's ecological interpretation of the ideas of the young Marx, and what *is* apparent in the writings of the young Marx and in other versions of humanist eco-Marxism, is a notion of human freedom that is irredeemably anthropocentric. This notion of human freedom takes its meaning from Marx's distinction between freedom and necessity.

It will be recalled that the young Marx maintained that humans realize their "essence" through unalienated labor and that the distinctive characteristic of humans was that "they produce universally, that is, produce even when free from physical need *and only truly produce in freedom thereof.*"[59] This same distinction also runs through the writings of the mature Marx. For example, in *Capital* Marx wrote:

> The realm of freedom actually begins only where labor which is determined by necessity and mundane considerations ceases; thus in the nature of things it lies beyond the sphere of actual material production.[60]

According to Marcuse and Gorz (both of whom build on this freedom/necessity distinction), the more we have mastered necessity, the more we can become truly free and realize our individuality through creative leisure, the sciences and the arts, convivial activity, and the like. According to Marcuse (who also drew on Freud's theory of human instincts), freedom lies in eros and play, not labor, for labor presupposes the suppression of instincts and the conquering of desire. In other words, the problem lies in the fact that social necessity demands that humans must always labor. Unlike Freud, however, Marcuse argued that, in today's society, scarcity (which gives rise to "basic repression") is not so much a brute fact as a consequence of a specific social organization that is sustained to secure the privileged position of powerful groups and individuals—a state of affairs that has led to "surplus repression."[61] Marcuse observed that, paradoxically, the very technological achievements of "repressive civilization" (which he considered to be dominated by the "performance principle"—the prevailing historical form of the "reality principle") have created the preconditions for the gradual abolition of repression. That is, breaking down these social relations of domination would enable the forces of production to be pressed into the service of "genuine need" by liberating humans from toil, thereby creating a post-scarcity and hence "nonrepressive civilization." Like the young Marx (and contra the mature Marx), Marcuse believed that both labor and scarcity could be abolished in this way rather than simply diminished.

The problem with this humanist eco-Marxist project of overcoming human alienation is that "true" human freedom and embeddedness in nature are posited as *inversely related*.[62] That is, the kind of freedom pursued by humanist eco-Marxists necessarily requires the subjugation of external nature (through labor's extension, technology) so that humans may ultimately become fully sovereign and answerable only to themselves, as opposed to being dependent on, and "held down" by, the limitations and inconveniences of nonhuman nature. Nonhuman nature remains, as Benton observes, "an external, threatening and constraining power...to be overcome in the course of a long-drawn-out historical process of collective transformation."[63]

The ultimate purport, then, of Marx's notion of "the resurrection of nature" in *The Economic and Philosophical Manuscripts of 1844* is not a nonanthropocentric socialist stewardship of "our inorganic body," as Lee would have us believe, but rather the further subjugation of the nonhuman world.[64] If Lee's humanist eco-Marxism is to remain recognizable as Marxism, then it must accept the anthropocentric implications of Marx's particular notion of freedom, which takes its meaning from the problematic freedom/necessity distinction.

In any event, some ecocentric critics have argued that even Lee's apparently benign ecocentric interpretation of the ideas of the young Marx—an interpretation that focuses on the theme of alienation—nonetheless serves to legitimize (albeit unwittingly) the domination of the nonhuman world. According to Lee, the overcoming of our alienation from nature is understood as the outcome of a dialectical struggle (sometimes referred to as a "metabolic interaction") between Subject (the laborer) and Object (external nature, the material to be transformed). According to this view, for nature to be recognized as our "body," a familiar and extended part of us rather than something "other," there must be a mutual interpenetration of both spheres through the activity of labor resulting in the mutual transformation of both— thus arriving at the much heralded "humanization of nature" and "naturalization of humanity." Ecocentric critics have suggested that this superficially attractive version of overcoming the human/nature dichotomy is yet another form of domination couched in the language of human self-realization.[65] According to Val Routley, Marx's early view of nature as our body, our creation and our expression

> can usefully be seen as the product of Marx's well-known transposition of God's features and role in the Hegelian system of thought onto man.... Thus, Marx's theory represents an extreme form of the placing of man in the role previously attributed to God, a transposition so characteristic of Enlightenment thought.[66]

The upshot of humanist eco-Marxism (Lee's version included) is that *the unity of humans with nature is achieved by making it our artifact, by totally domesticating it*. We have thus returned full circle to the orthodox solution to the environmental crisis (albeit couched in different language), to that stage of human development where nature is totally mastered through the power of associated individuals. In Lee's own words, this "unity" would enable us to live "in consciousness that each of us is identical with each other and with nature, and exploitation of men and nature would cease."[67]

From an ecocentric perspective, however, harmonizing our relationship with the rest of nature does not mean obliterating or humanizing what is "other" or not-human in nature. Rather, it means *identifying* with it (not making it identical to us) in a way that involves the recognition of the *relative autonomy* and unique mode of being of the myriad life-forms that make up the nonhuman world. To ecocentric theorists, freedom or self-determination is recognized as a legitimate entitlement of *both* human and nonhuman life-forms. As we saw in part 1, the goal of an ecocentric political theory is "emancipation writ large"—the maximization of the freedom of *all* entities to unfold or develop in their own ways. Moreover, such freedom or self-determination is understood in relational terms (both socially and ecologically), in that the development of any relatively autonomous parts of a larger system (e.g., an ecosystem or the ecosphere) is inextricably tied to its relationship with the development of other relatively autonomous parts of that system as well as the development of the system itself (i.e., the whole). Ecocentric theorists argue that whereas the flourishing of human life and culture is quite compatible with a human lifestyle based on low material and energy throughput, the flourishing of nonhuman life *requires* such a human lifestyle. In order to meet this requirement we need to live and experience ourselves as but one component of, and more or less keep pace with, the basic cycles and processes of nature rather than seek to totally transcend the nonhuman world by removing all of its inconveniences and thereby obliterating its "otherness."

To be sure, Lee's reinterpretation of the "mastery" of nature to mean "rational harmony with nature" is more ecologically grounded than that of orthodox eco-Marxists, who would encourage the development of all manner of synthetic substitutes for "nature's bounty" so as to avoid remaining "tethered" to the cycles of nature. Yet this latter kind of outcome is logically entailed in the quest for freedom according to the young Marx. As we have seen, if true freedom is understood to be inversely related to our embeddedness in nature, then the realization of that freedom necessarily requires that we seek to increase our control over, and reduce our dependence on, ecological cycles. The upshot is that nature, although redefined as "our body," must be thoroughly tamed and made subservient to human ends.

Of course, the kind of "true human freedom" promised by humanist eco-Marxists is an attractive and familiar interpretation of "freedom," made all the more so in the light of the general bifurcation between work and leisure in modern society. Yet, from an ecocentric perspective, this eco-Marxist coupling of necessity/freedom and work/leisure is objectionable in two important respects. First, it is based on the anthropocentric Marxist "differential imperative" of *homo faber*. It will be recalled from our discussion in chapter 3 that arguing by way of the "differential imperative" means selecting certain characteristics that are believed to be special to humans vis-à-vis other species as the measure of both human virtue and human superiority over other species. According to Marx, to be fully human and truly free, humans must maximize what he believed made us different from the rest of nature, namely, our ability to self-consciously act upon and transform the external world and thereby augment our own powers.[68] Yet, like so many anthropocentric assumptions, Marx's putative human/nonhuman opposition is based on an erroneous understanding of ecological reality. As Benton puts it in an extended critique of this opposition:

> For his intellectual purposes, Marx exaggerates both the fixity and limitedness of scope in the activity of other animals, and the flexibility and universality of scope of human activity upon the environment.[69]

Second, the eco-Marxist distinction between freedom and necessity reifies nonessential activities as the means to true individual fulfilment while downgrading as lowly or "animal-like" many life-sustaining activities that can be potentially more fulfilling if approached and organized differently. (By life-sustaining activities, I am referring to those fundamental human activities necessary for survival and physical and psychological health such as growing and preparing food, constructing and maintaining shelter, nurturing and teaching the young, and caring for the infirm and the elderly.) The result is that culture and self-expression are made the complete antithesis of necessary material labor. This is because the general reduction of necessary labor time is seen to provide the foundation for the "true freedom" to be experienced by all humans, a freedom to be "purchased" via ongoing technological developments designed to "relieve" humans from concerning themselves with those burdensome tasks that have limited the development of present and prior generations of humans and which are seen to remain forever the lot of the rest of the animal world.[70] Marx's distinction between freedom and necessity thus creates a dualism not only between humans and nonhuman nature but also within human nature between common, lowly animal functions, powers, and needs and *sui generis,* higher human functions, powers, and needs.[71] As we saw in chapter 3, Simone de Beauvoir has drawn attention to a similar kind of contrasting valuation between nature and culture and between the self-limiting work of women and the self-transcending work of men. The former is treated as private, mundane, and concerned with the regeneration and repetition of life while the latter is regarded as public, worthy, and concerned with transcending life by reshaping the future through technology and symbols.

The ecocentric objection to the post-scarcity utopia of humanist eco-Marxism is that it would cultivate a type of human who, as Val Routley has observed, through science and technology, is thoroughly insulated from, and in control of, the cycles of nature and the myriad of other nonhuman life-forms. Indeed, it is hard to see how the overcoming of human alienation from nature is to be achieved in such a utopia, when humans are to be so thoroughly insulated and removed from their biological roots. In effect, the price of overcoming alienation in the workplace is alienation from nonhuman nature. Moreover, as Bookchin has argued, according to the Marxist view of freedom, class society and authoritarian social relations will remain unavoidable for so long as the mode of production fails to provide a sufficient material abundance for everybody to enjoy the realm of "true freedom." Until that time, the "realm of necessity" will become "a realm of command and obedience, of ruler and ruled."[72] That is, the domination of people and the rest of nature under a rationalized capitalism remains the precondition for the achievement of a distant and continually postponed socialist freedom.

It is clear that Marcuse's version of humanist eco-Marxism is more firmly grounded in the philosophical ideas of the young Marx than Lee's version. Indeed, it is noteworthy that Tolman has rejected Lee's selective reading of Marx's early writings for being decidedly un-Marxist. According to Tolman, Lee's basic argument that we ought to see ourselves as stewards of the environment, which is to be seen as part of our "inorganic" body, is a serious distortion of Marx's true position, since Marx ultimately rejected the "early ideas" drawn upon by Lee as abstract and idealist.[73] We have already seen that the mature Marx had come to the view that the antagonistic struggle between humanity and nature would never be completely resolved, that is, nature would never be fully "resurrected" since labor could never be totally abolished and, accordingly, there would always be some parts of nature that remained untransformed and alien. For his part, Tolman makes no apologies in declaring the incompatibility between Marxism and ecocentrism. On this score, Tolman seems to have a clearer perception than Lee of the long term technological and ecological implications of Marxism.

From an ecocentric perspective, then, the Marxist dichotomy between freedom and necessity must be transcended if we are to allow the mutual unfolding of both human and nonhuman life. In particular, the view (strongly endorsed by Gorz in particular) that even unalienated, self-managed material labor is a "lower" form of freedom than unnecessary labor and/or leisure activity must be rejected. Rather, the emphasis must turn to exploring the many ways in which basic needs may be met and necessary and life-sustaining work performed in a manner that is personally, aesthetically, and intellectually satisfying and *not* environmentally damaging.[74] Individuality, self-expression, and rounded human development will then be able to be realized *through* socially useful work as well as through other kinds of activity. (As I show in my discussion of ecoanarchism is chapter 7, one kind of setting in which rounded human development is possible is a self-managed, cooperative community.) Moreover, contrary to Gorz's presumption, the enjoyment of "true freedom" as leisure need not be dependent on high technology or a high energy and material throughput. As Marshall Sahlins has argued in *Stone Age Economics,* the enjoyment of affluence (i.e., interpreted here as ample creative leisure rather than as an abundance of material goods) by the members of a particular society is not necessarily dependent on that society "mastering necessity" by conquering nature through advanced technology and a high energy consumption.[75] Quite the contrary, his book is an important illustration of the maxim "want not, lack not" and a challenge to what he calls modern culture's "shrine to the Unattainable: Infinite Needs."[76] From an ecocentric perspective, such creative leisure is best procured through the critical revision and simplification of human needs and the development of tools and goods appropriate to those revised needs rather than through the systematic replacement of human labor by energy intensive machines.

Beyond Marxism

This chapter has sought to show that an ecocentric perspective cannot be wrested out of Marxism, whether orthodox or humanist, without seriously distorting Marx's own theoretical concepts. As John Clark has put it, "to develop the submerged ecological dimension of Marx would mean the negation of key aspects of his philosophy of history, his theory of human nature, and his view of social transformation."[77] This explains why nonanthropocentric Green theorists have chosen not to develop their ideas within a Marxist framework and instead have sought guidance from other traditions of political thought such as utopian socialism, communal anarchism, and feminism. Far from providing a theoretical touchstone, Isaac Balbus has argued that the time has come for ecologically oriented political theorists to do to the ideas of Marx what he did to the ideas of the bourgeois thinkers he contested, namely, explain their origins in order to reveal their historical limits.[78] In this respect, Hwa Yol Jung has provided three succinct reasons why ecologically oriented theorists should abandon the ideas of Marx:

> first, he was too Hegelian to realize that the gain of "History" (or Humanity) is the loss of "Nature"; second, he was influenced by the English classical labor theory of value which undergirds his conception of man as *homo faber;* and third, he was a victim of the untamed optimism of the Enlightenment for Humanity's future progress.[79]

To be sure, Marx's conception of freedom was more comprehensive than the liberal concept of freedom that he called to account. As Booth has neatly put it, whereas liberalism had grasped "one form of unfreedom, coercion, or the arbitrary rule of one will over another, which was the dominant form in pre-capitalist societies," Marxism recognized another form, namely, the silent and objective compulsion of the economic laws of capitalism.[80] As ecocentric theorists have shown, however, neither liberalism nor Marxism has acknowledged the unfreedom of the nonhuman world under *industrialism*.[81] To acknowledge the particular kinds of unfreedom exposed by Marx, as ecocentric theorists do (given their general concern for "emancipation writ large"), does not also require an acceptance of Marx's notion of freedom. Quite the contrary. I have sought to show that Marx's notion of freedom as *mastery* achieved through struggle, as the subjugation of the external world through labor and its extension—technology, as the conquering of mysterious or hostile forces and the overcoming of all constraints, can be achieved only at the expense of the nonhuman world.

In this chapter, my discussion has generally been confined to those strands of eco-Marxism that have sought inspiration from Marx's own texts. However, there is one more relatively distinct subset of humanist Marxist thought, namely, the Critical Theory of the Frankfurt School, that warrants the special attention of ecocentric theorists on account of its innovative critique of instrumental reason—a subject largely left unexplored by the theorists discussed in this chapter (with the exception, of course, of Marcuse, who is himself a Critical Theorist).

CHAPTER 5

The Failed Promise of Critical Theory

INTRODUCTION

The Critical Theory developed by the members of the Frankfurt Institute of Social Research ("the Frankfurt School") has revised the humanist Marxist heritage in ways that directly address the wider emancipatory concerns of ecocentric theorists.[1] In particular, Critical Theorists have laid down a direct challenge to the Marxist idea that "true freedom" lies beyond socially necessary labor. They have argued that the more we try to "master necessity" through the increasing application of instrumental reason to all spheres of life, the *less* free we will become.

Critical Theory represents an important break with orthodox Marxism— a break that was undertaken in order to understand, among other things, why Marx's original emancipatory promise had not been fulfilled. Like many other strands of Western Marxism, Critical Theory turned away from the scientism and historical materialism of orthodox Marxism. In the case of the Frankfurt School, however, it was not through a critique of political economy but rather through a critique of culture, scientism, and instrumental reason that Marxist debates were entered. One of the enduring contributions of the first generation of Frankfurt School theorists (notably, Max Horkheimer, Theodor Adorno, and Herbert Marcuse) was to show that there are different levels and dimensions of domination and exploitation *beyond* the economic sphere and that the former are no less important than the latter. The most radical theoretical innovation concerning this broader understanding of domination came from the early Frankfurt School theorists' critical examination of the relationship between humanity and nature. This resulted in a fundamental challenge to the orthodox Marxist view concerning the progressive march of history, which had emphasized the liberatory potential of the increasing mastery of nature through the development of the productive forces. Far from welcoming these developments as marking the "ascent of man from the kingdom of necessity to the kingdom of freedom" (to borrow Engels's phrase), Horkheimer, Adorno, and Marcuse saw them in essentially negative terms, as giving rise to the domination of both "outer" and "inner" nature.[2] These early Frankfurt School theorists regarded the rationalization process set in train by the Enlightenment as a "negative dialectics." This was reflected, on the one hand, in the apprehension and conversion of nonhuman nature into resources for production or objects of scientific inquiry (including animal experimentation) and, on the other hand, in the repression of humanity's joyful and spontaneous instincts brought about through a repressive social division of labor and a repressive division of the human psyche. Hence their quest for a human "reconciliation" with nature. Instrumental or "purposive" rationality—that branch of human reason that is concerned with determining the most efficient means of realizing pregiven goals and which accordingly apprehends only the instrumental (i.e., use) value of phenomena—should not, they argued, become the exemplar of rationality for society. Human happiness would not come about simply by improving our techniques of social administration, by treating society and nature as subject to blind, immutable laws that could be manipulated by a technocratic elite.

The early Frankfurt School's critique of instrumental rationality has been carried forward and extensively revised by Jürgen Habermas, who has sought to show, among other things, how political decision making has been increasingly reduced to pragmatic instrumentality, which serves the capitalist and bureaucratic system while "colonizing the life-world."[3] According to Habermas, this "scientization of politics" has resulted in the lay public ceding ever greater areas of system-steering decision making to technocratic elites.

All of these themes have a significant bearing on the Green critique of industrialism, modern technology, and bureaucracy, and the Green commitment to grassroots democracy. Yet Critical Theory has not had a major direct influence in shaping the theory and practice of the Green movement in the 1980s, whether in West Germany or elsewhere.[4] We saw in chapter 1, however, that the ideas of Marcuse and Habermas did have a significant impact on the thinking of the New Left in the 1960s and early 1970s and that the general "participatory" theme that characterized that era has remained an enduring thread in the emancipatory stream of ecopolitical thought. Yet this legacy is largely an indirect one. Of course, there are some emancipatory theorists who have drawn upon Habermas's social and political theory in articulating and explaining some aspects of the Green critique of advanced industrial society.[5] However, this can be contrasted with the much greater general influence of post-Marxist Green theorists such as Murray Bookchin, Theodore Roszak, and Rudolf Bahro and non-Marxist Green theories such as bioregionalism, deep/transpersonal ecology, and ecofeminism—a comparison that further underscores the distance Green theory has had to travel away from the basic corpus of Marxism and neo-Marxism in order to find a comfortable "theoretical home."

It is important to understand why Critical Theory has not had a greater direct impact on Green political theory and practice given that two of its central problems—the triumph of instrumental reason and the domination of nature—might have served as useful theoretical starting points for the Green critique of industrial society. This possibility was indeed a likely one when it is remembered that both the Frankfurt School and Green theorists acknowledge the dwindling revolutionary potential of the working class (owing to its integration into the capitalist order); both are critical of totalitarianism, instrumental rationality, mass culture, and consumerism; and both have strong German connections. Why did these two currents of thought not come together?

There are many possible explanations as to why Critical Theory has not been more influential. One might note, for example, the early Frankfurt School's pessimistic outlook (particularly that of Adorno and Horkheimer), its ambivalence toward nature romanticism (acquired in part from its critical inquiry into Nazism), its rarefied language, its distance from the imperfect world of day-to-day political struggles (Marcuse being an important exception here), and its increasing preoccupation with theory rather than praxis (despite its original project of uniting the two). Yet a more fundamental explanation lies in the direction in which Critical Theory has developed since the 1960s in the hands of Jürgen Habermas, who has, by and large, remained preoccupied with and allied to the fortunes of democratic socialism (represented by the Social Democratic Party in West Germany) rather than the fledgling Green movement and its parliamentary representatives.[6] Of course, the Green movement has not escaped Habermas's attention. However, he has tended to approach the movement more as an indicator of the motivational and legitimacy problems in advanced capitalist societies rather than as the historic bearer of emancipatory ideas (this is to be contrasted with Marcuse, who embraced the activities of new social movements).[7] Habermas has analyzed the emergence of new social movements and Green concerns as a grassroots "resistance to tendencies to colonize the life-world."[8] With the exception of the women's movement (which Habermas does consider to be emancipatory), these new social movements (e.g., ecology, antinuclear) are seen as essentially *defensive* in character.[9] While acknowledging the ecological and bureaucratic problems identified by these movements, Habermas regards their proposals to develop counterinstitutions and "liberated areas" from *within the life-world* as essentially unrealistic. What is required, he has argued, are "technical and economic solutions that must, in turn, be planned globally and implemented by administrative means."[10] Yet as Anthony Giddens has pointedly observed, if the pathologies of advanced industrialism are the result of the triumph of purposive rationality, how can the life-world be defended against the encroachments of bureaucratic and economic steering mechanisms without transforming those very mechanisms?[11] In defending the revolution-

ary potential of new social movements, Murray Bookchin has accused Habermas of intellectualizing new social movements "to a point where they are simply incoherent, indeed, atavistic."[12] According to Bookchin, Habermas has no
sense of the potentiality of new social movements.

Yet Habermas's general aloofness from the Green movement (most
notably, its radical ecocentric stream) goes much deeper than this. It may be
traced to Habermas's theoretical break with the "negative dialectics" of the
early Frankfurt School theorists and with their utopian goal of a "reconciliation with nature." Habermas has argued that such a utopian goal is neither
necessary nor desirable for human emancipation. Instead, he has welcomed
the rationalization process set in train by the Enlightenment as a *positive*
rather than negative development. This chapter will be primarily concerned
to locate this theoretical break and outline the broad contours of the subsequent development of Habermas's social and political theory in order to
identify what I take to be the major theoretical stumbling blocks in Habermas's *oeuvre*. This will help to explain, on the one hand, why Habermas
regards the radical ecology movement as defensive and "neo-romantic" and,
on the other hand, why ecocentric theorists would regard many of Habermas's theoretical categories as unnecessarily rigid and anthropocentric.

In contrast, a central theme of the early Frankfurt School theorists,
namely, the hope for a reconciliation of the negative dialectics of Enlightenment that would liberate both human and nonhuman nature, speaks directly
to ecocentric concerns. While Adorno and Horkheimer were pessimistic as to
the prospect of such a reconciliation ever occurring, Marcuse remained hope-
ful of the possibility that a "new science" might be developed, based on a
more expressive and empathic relationship to the nonhuman world. This
stands in stark contrast to Habermas's position—that science and technology
can know nature only in instrumental terms since that is the only way in
which it can be effective in terms of securing our survival as a species.
Unlike Habermas, Marcuse believed that a qualitatively different society
might produce a qualitatively different science and technology. Ultimately,
however, Marcuse's notion of a "new science" remained vague and undeveloped and, in any event, was finally overshadowed—indeed contradicted—by
his overriding concern for the emancipation of the human senses and the
freeing up of the instinctual drives of the individual. As we saw in the previous chapter, this required nothing short of the total abolition of necessary
labor and the rational mastery of nature, a feat that could be achieved *only* by
advanced technology and widespread automation.

The Legacy of Horkheimer, Adorno, and Marcuse

The contributions of Horkheimer and Adorno in the 1940s, and Marcuse in
the 1950s and 1960s, contain a number of theoretical insights that foreshad-

owed the ecological critique of industrial society that was to develop from the late 1960s.[13] Indeed, these insights might have provided a useful starting point for ecocentric theorists by providing a potential theoretical linkage between the domination of the human and nonhuman worlds. By drawing back from the preoccupation with class conflict as the "motor of history" and examining instead the conflict between humans and the rest of nature, Horkheimer and Adorno developed a critique that sought to transcend the socialist preoccupation with questions concerning the control and distribution of the fruits of the industrial order. In short, they replaced the critique of political economy with a critique of technological civilization. As Martin Jay has observed, they found a conflict whose origins predated capitalism and whose continuation (and probable intensification) appeared likely to survive the demise of capitalism.[14] Domination was recognized as increasingly assuming a range of noneconomic guises, including the subjugation of women and cruelty to animals—matters that had been overlooked by most orthodox Marxists.[15] The Frankfurt School also criticized Marxism for reifying nature as little more than raw material for exploitation, thereby foreshadowing aspects of the more recent ecocentric critique of Marxism. Horkheimer and Adorno argued that this stemmed from the uncritical way in which Marxism had inherited and perpetuated the paradoxes of the Enlightenment tradition—their central target. In this respect, Marxism was regarded as no different from liberal capitalism.

Horkheimer and Adorno's contribution was essentially conducted in the form of a philosophical critique of reason. Their goal was to rescue reason in such a way as to bring instrumental reason under the control of what they referred to as "objective" or "critical" reason. By the latter, Adorno and Horkheimer meant that synthetic faculty of mind that engages in critical reflection and goes beyond mere appearances to a deeper reality in order to reconcile the contradictions between reality and appearance. This was to be contrasted with "instrumental reason," that one-sided faculty of mind that structures the phenomenal world in a commonsensical, functional way and is concerned with efficient and effective adaptation, with means, not ends. The Frankfurt School theorists sought to defend reason from attacks on both sides, that is, from those who reacted against the rigidity of abstract rationalism (e.g., the romanticists) and from those who asserted the epistemological supremacy of the methods of the natural sciences (i.e., the positivists). The task of Critical Theory was to foster a mutual critique and reconciliation of these two forms of reason. In particular, reason was hailed by Marcuse as an essential "critical tribunal" that was the core of any progressive social theory; it lay at the root of Critical Theory's utopian impulse.[16]

According to Horkheimer and Adorno, the Age of Enlightenment had ushered in the progressive replacement of tradition, myth, and superstition with reason, but it did so at a price. The high ideals of that period had become grossly distorted as a result of the ascendancy of instrumental reason over critical reason, a process that Max Weber decried as simultaneously leading to the rationalization *and* disenchantment of the world. The result was an inflated sense of human self-importance and a quest to dominate nature. Horkheimer and Adorno argued that this overemphasis on human self-importance and sovereignty led, paradoxically, to a loss of freedom. This arose, they maintained, because the instrumental manipulation of nature that flowed from the anthropocentric view that humans were the measure of all things and the masters of nature inevitably gave rise to the objectification and manipulation of humans:

> Men pay for the increase in their power with alienation from that over which they exercise their power. Enlightenment behaves toward things as a dictator toward men. He knows them in so far as he can manipulate them. The man of science knows things in so far as he can make them. In this way their potentiality is turned to his own ends. In the metamorphosis the nature of things, as the substratum of domination, is revealed as always the same. This identity constitutes the unity of nature.[17]

The first generation of Critical Theorists also argued that the "rational" domination of outer nature necessitated a similar domination of inner nature by means of the repression and renunciation of the instinctual, aesthetic, and expressive aspects of our being. Indeed, this was seen to give rise to the paradox that lay at the heart of the growth of reason. The attempt to create a free society of autonomous individuals via the domination of outer nature was self-vitiating because this very process also distorted the subjective conditions necessary for the realization of that freedom.[18] The more we seek material expansion in our quest for freedom from traditional and natural constraints, the more we become distorted psychologically as we deny those aspects of our own nature that are incompatible with instrumental reason. As Alford has observed, Horkheimer and Adorno condemned "not merely science but the Western intellectual tradition that understands reason as effective adaptation."[19] Whereas Weber had described the process of rationalization as resulting in the disenchantment of the world, Horkheimer and Adorno described it as resulting in the "revenge of nature." This was reflected in the gradual undermining of our biological support system and, more significantly, in a new kind of repression of the human psyche. Such "psychic repression" was offered as an explanation for the modern individual's blind susceptibility, during times of social and economic crisis, to follow those demagogues (Hitler being the prime example) who offer the alienated individual a sense of meaning and belonging. From a Critical Theory perspective, then, just as the totalitarianism of Nazism was premised on the will to *engineer* social problems out of existence, the bureaucratic state and corporate capitalism may be seen as seeking to *engineer* ecological problems out of existence.

Adorno, Horkheimer, and Marcuse longed for "the resurrection of nature"—a new kind of mediation between society and the natural world. Whitebook has described this resurrection as referring to "the transformation of our relation to and knowledge of nature such that nature would once again be taken as purposeful, meaningful or as possessing value."[20] This did not mean a nostalgic regress into primitive animism or pre-Enlightenment mythologies that sacrificed critical reason—the phenomenon of Nazism demonstrated the dangers of such a simplistic solution. Rather, their utopia required the *integrated* recapture of the past. This involved remembering rather than obliterating the experiences and ways of being of earlier human cultures and realizing that the modern rationalization process and the increasing differentiation of knowledge (particularly the factual, the normative, and the expressive) has been both a learning *and* unlearning process. What was needed, Adorno, Horkheimer, and Marcuse believed, was a new harmonization of our rational faculties and our sensuous nature.[21]

Yet Adorno and Horkheimer recognized that their utopia was very much against the grain of history. Unlike Marx, they stressed the *radical discontinuity* between the march of history and the liberated society they would like to see. As we saw, this sprang from the lack of a revolutionary subject that would be able to usher in the reconciliation of humanity with inner and outer nature. After all, how could there be a revolutionary subject when the individual in mass society had undergone such psychological distortion and was no longer autonomous? Accordingly, they were unable to develop a revolutionary praxis to further their somewhat vague utopian dream. However, they insisted that the utopian impulse that fuelled that dream, although never fully realizable, must be maintained as providing an essential source of critical distance that guarded against any passive surrender to the status quo.

Although Marcuse explored the same negative dialectics as Adorno and Horkheimer, he reached a more optimistic conclusion concerning the likelihood of a revolutionary praxis developing. In particular, he saw the counterculture and student movements of the 1960s and early 1970s as developing a more expressive relationship to nature that was cooperative, aesthetic—even erotic. Here, he suggested, were the seeds of a new movement that could expose the ideological functions of instrumental rationality and mount a far-reaching challenge to the "false" needs generated by modern consumer society that had dulled the individual's capacity for critical reflection.[22] Marcuse saw aesthetic needs as a subversive force because they enabled things to be seen and appreciated *in their own right*.[23] Indeed, he argued that the emancipation of the senses and the release of instinctual needs was a prerequisite to the liberation of nature (both internal and external). In the case of the former, this meant the liberation of our primary impulses and aesthetic senses. In the case of the latter, it meant the overcoming of our incessant struggle with our environment and the recovery of the "life-enhancing forces in nature, the sensuous aesthetic qualities which are foreign to a life wasted in unending competitive performance."[24]

Marcuse also advanced the provocative argument that this kind of "sensuous perception" might form the epistemological basis of a new science that would overcome the one-dimensionality of instrumental reason that he believed underpinned modern science. Under a new science, Marcuse envisaged that knowledge might become a source of pleasure rather than the means of extending human control. The natural world would be perceived and responded to in an open, more passive, and more receptive way and be guided by the object of study (rather than by human purposes). Such a new science might also reveal previously undisclosed aspects of nature that could inspire and guide human conduct.[25] This was to be contrasted with modern "Galilean" science, which Marcuse saw as "the 'methodology' of a pre-given historical reality within whose universe it moves"; it reflects an interest in experiencing, comprehending, and shaping "the world in terms of calculable, predictable relationships among exactly identifiable units. In this project, universal quantifiability is a prerequisite for the *domination* of nature."[26]

Habermas has taken issue with Marcuse, claiming that it is logically impossible to imagine that a new science could be developed that would overcome the manipulative and domineering attitude toward nature characteristic of modern science.[27] There are certainly passages in Marcuse's *One Dimensional Man* that suggest that it is the scientific method itself that has ultimately led to the domination of humans and that therefore a change in the very method of scientific inquiry is necessary to usher in a liberated society.[28] Against Habermas's interpretation, however, William Leiss has argued that these are isolated, inconsistent passages that run contrary to the main line of Marcuse's argument, which is that the problem is not with science or instrumental rationality per se but "with the repressive social institutions which exploit the achievements of that rationality to preserve unjust relationships."[29]

Yet these inconsistencies in Marcuse's discussion of the relationship between science and liberation do not appear to be resolvable either way. Indeed, it is possible to discern a third position that lies somewhere between Habermas's and Leiss's interpretations (although it is closer to Leiss's): that the fault lies neither with science nor instrumental rationality per se nor repressive social institutions per se but rather with the instrumental and anthropocentric character of the modern worldview. In *One Dimensional Man,* Marcuse was concerned to highlight the inextricable interrelationship between science and society. He conceded that pure as distinct from applied science "does not project particular practical goals nor particular forms of domination," but it does proceed in a certain universe of discourse and cannot transcend that discourse.[30] According to Marcuse, scientific rationality was in itself, in its very abstractness and purity, operational in as much as it developed under an instrumental horizon.... This interpretation would tie the scientific project (method and theory), *prior* to all application and utilization, to a specific societal project, and would see the tie precisely in the inner form of scientific rationality, i.e., the functional character of its concepts.[31]

It is clear that Marcuse regarded the scientific method as being dependent on a pre-established universe of ends, in which and *for* which it has developed.[32] It follows, as he points out in *Counterrevolution and Revolt,* that:

A free society may well have a very different a priori and a very different object; the development of the scientific concepts may be grounded in an experience of nature as a totality of life to be protected and "cultivated," and technology would apply this science to the reconstruction of the environment of life.[33]

Marcuse's point is a very general one: that a new or liberatory science can only be inaugurated by a liberatory society. It would be a "new" science because it would serve a new preestablished universe of ends, including a qualitatively new relationship between humans and the rest of nature. This third interpretation is much closer to Leiss's interpretation than Habermas's since it argues that we must reorder social relations before we reorder science if we wish to "resurrect" nature. Only then would we be able to cultivate a liberatory rather than a repressive mastery of nature.

Yet it is important to clarify what Marcuse meant by a "liberatory mastery of nature." As Alford has convincingly shown, Marcuse's new science appears as mere rhetoric when judged against the overall thrust of his writings.[34] As we saw in the previous chapter, Marcuse's principal Marxist reference point was the *Paris Manuscripts,* which Marcuse saw as providing the philosophical grounding for the realization of the emancipation of the senses and the reconciliation of nature. Moreover, his particular Marx/Freud synthesis was concerned to overcome repressive dominance, that is, the repression of the pleasure principle (the gratification of the instincts) by the reality principle (the need to transform and modify nature in order to survive, which is reflected in the work ethic and the growth of instrumental reason). Marcuse saw the reality principle as being culturally specific to an economy of scarcity. In capitalist society, the forces of production had developed to the point where scarcity (which gave rise to the "reality principle") need no longer be a permanent feature of human civilization. That is, the technical and productive apparatus was seen to be capable of meeting basic necessities with minimum toil so that there was no longer any basis for the repression of the instincts via the dominance of the work ethic. The continuance of this ethic must be seen as "surplus repression," which Marcuse maintained was secured, inter alia, by the manipulation of "false" consumer needs.[35] Marcuse ultimately wished to reap the full benefits promised by mainstream science, namely, a world where humans would be spared the drudgery of labor and be free to experience "eros and peace."

However, the necessary quid pro quo for the reassertion of the pleasure principle over the reality principle was that the nonhuman world would continue to be sacrificed in the name of human liberation. Marcuse shared Marx's notion of two mutually exclusive realms of freedom and necessity and, like Marx, he believed that "true freedom" lay beyond the realm of labor. Accordingly, total automation, made possible by scientific and technological progress, was essential on the ground that necessary labor was regarded as inherently unfree and burdensome. It demanded that humans subordinate their desires and expressive instincts to the requirements of the "objective situation" (i.e., economic laws, the market, and the need to make a livelihood). I have already discussed the limitations of humanist eco-Marxism in the previous chapter and need not repeat all of these criticisms here. It will suffice simply to emphasize that socialist stewardship under humanist eco-Marxism would usher in a "reconciliation with nature" of a kind that would see to the total domestication or humanization of the nonhuman world. As Malinovich has observed, "for Marcuse the concept of the 'development of human potentiality for its own sake' became *the* ultimate socialist value."[36] In Marcuse's own words, the emancipation of the human senses under a humanistic socialism would enable

> "the human appropriation of nature," i.e., through the transformation of nature into an environment (medium) for the human being as "species being"; free to develop the specifically human faculties: the creative, aesthetic faculties.[37]

Despite his intriguing discussion of the notion of a new, nondomineering science, then, Marcuse's major preoccupation with human self-expression, gratification, and the free play of the senses ultimately overshadowed his concern for the liberation of nonhuman nature. Any nonanthropocentric gloss that Marcuse may have placed on Marx's *Paris Manuscripts* must be read down in this context. Nonetheless, Marcuse's "ecocentric moments" (i.e., his discussion of a qualitatively different science and society that approach the nonhuman world as a partner rather than as an object of manipulation) serve as a useful foil to Habermas's more limited conceptualization of the scientific project.

HABERMASIAN REVISIONS

Habermas has carried forward but extensively revised the early Frankfurt School's critique of instrumental reason in advanced industrial society. He

has argued that the advance of instrumental reason has led to the "scientization of politics," that process whereby social and environmental problems are
increasingly posed as technical problems requiring technical solutions by
experts rather than as political problems that need to be addressed, first and
foremost, by an informed and rational citizenry.[38] According to Habermas,
this is part of a larger process that has been taking place over the last two
hundred years—beginning with Hobbes—involving the gradual demise of
the classical doctrine of politics (which had entailed the cultivation of "practical wisdom") and the emergence of specialized social sciences that emulate
the methodology of the natural sciences.[39] The result is that the achievement
of a rational, democratic consensus by an informed citizenry concerning
societal goals is being increasingly subverted by a technical discussion by a
minority of experts concerning means (based on presupposed ends, namely,
economic growth, the expansion of the bureaucratic-technical apparatus, and
the domination of human and nonhuman nature). This has led to the depoliti-
cization, manipulation, and unacknowledged domination of the majority of
the population by a technical and bureaucratic elite and the concomitant
withering of the "public sphere" (culminating in the decline of parliament as
a meaningful forum for debate). According to Habermas, reason has lost the
critical force it once had. It is now degraded to mean only instrumental reason with the result that "the industrially most advanced societies seem to
approximate the model of behavioral control steered by external stimuli
rather than guided by norms."[40]

Unlike his Frankfurt School predecessors, however, Habermas does not
argue for the "resurrection of a fallen nature," that is, a healing of the rift
between humanity and nonhuman nature that has been brought about by the
rationalization process. Nor does he accept the need for a "new science."
Instead, Habermas has taken a different path by locating instrumental reason
within a larger and more comprehensive theory of rationality. He has criticized Adorno, Horkheimer, and Marcuse's central thesis—that the domination of "external" nature leads inexorably to the domination of "internal"
(i.e., human) nature—and has argued that the proper mastery of "external"
and "internal" nature does not follow the same logic of instrumental rationality. Habermas has posited instead a dualistic framework whereby the logic of
instrumental rationality governs our dealings with the nonhuman world and
the logic of communicative rationality governs interaction between human
subjects. According to Habermas, the former necessarily aims at manipulation, control, and reification whereas in the case of the latter manipulation,
control, and reification are possible but pathological outcomes (the proper
telos is autonomy, individuation, and socialization).[41]

According to Habermas, Marcuse's new science, which seeks to
approach nature as a partner rather than as an object of technical control,
confuses two different structures of action, namely, purposive-rational action

(the project of labor) and symbolic interaction or communication (the project of language).[42] Labor and its extension—technology—are seen as forming an indispensable part of the survival project of the human species as a whole. Science and technology, according to Habermas, are determined by the objective character of human labor, which is to wrestle with nonhuman nature (the bounty of which is all too scarce) in order to extract a livelihood. This relationship of labor (and technology) to nature is presented as having a "quasi-transcendental" status; that is, it is not historically relative but rather is a kind of biological drive that is rooted in the human species. It is on this ground that Habermas maintains that we can "know" nature (through our work and technology) only as an object of instrumental control. In contrast, communication (the project of language) is the concern of the historical-hermeneutic sciences, which are directed toward interpretive understanding, that is, "securing and expanding the possibilities of mutual and self understanding in the conduct of life."[43]

By separating labor and communication and grounding them in different cognitive interests (i.e., technical control and understanding), Habermas was able to reject the early Frankfurt School's pessimistic thesis that technical progress necessarily entailed moral regression and the distortion of the psychological conditions of emancipation. According to Habermas, what is needed is not a new science or new technologies but rather *undistorted communication*. As Alford has put it, "the goal of Habermas's project can be expressed in one sentence: to prevent social relations from becoming like our relations with the natural world."[44] The upshot is that the domination of non-human nature would continue as a legitimate project of the human species but that it would no longer entail the domination of humans in the way that the earlier Frankfurt School theorists had believed. Indeed, Habermas has hailed the progressive features of modernity and rejected what he regards as the "utopian excesses" of the early Frankfurt School theorists (such as their hope for the "reconciliation of nature"). The disenchantment of nature is accepted as the *necessary price* of human progress.

According to Habermas, the problems of advanced industrial societies do not stem from instrumental rationality per se but rather from the fact that instrumental rationality has not been accompanied or matched by a concomitant rationalization of social norms in the sphere of communication. By rationalization of social norms, Habermas means the establishment of a participatory democracy that provides the opportunity for undistorted communication and the achievement of a rational and universalistic normative consensus. These norms are to be found in what Habermas refers to as "the ideal speech situation."[45] Instrumental reason is presented as a specialized language abstracted out of ordinary communication, which, in turn, presupposes certain basic norms against which we may locate distortions in any given communication. As Habermas explains:

What raises us out of nature is the only thing whose nature we can know: *language.* Through its structure, autonomy and responsibility are posited for us. Our first sentence expresses unequivocally the intention of universal and unconstrained consensus. Taken together, autonomy and responsibility constitute the only Idea that we possess a priori in the sense of the philosophical tradition.[46]

Habermas's solution to the problem of the "scientization of politics," then, is not the reform of the logic of instrumental rationality per se but rather the reinvigoration of the public sphere (of "interaction") so that society can direct instrumental reason toward rationally justified ends.

Habermas's ideas and concerns have evolved considerably since he first took issue with the ideas of the early Frankfurt School. For example, instead of labor and interaction, he is now more likely to refer to "system" and "life-world," corresponding to purposive and communicative rationality respectively.[47] And in *The Theory of Communicative Action* he has moved away from a discussion of "quasi-transcendental" cognitive interests, preferring to ground Critical Theory in language or, more precisely, *communication* as distinct from epistemology.[48] In this recent project Habermas has outlined a theory of communication that is concerned to identify and clarify the conditions for undistorted human communication. A central purpose of this project has been to show that "the emancipatory critique does not rest on arbitrary norms that we 'choose'; rather it is grounded in the very structure of inter-subjective communicative competences."[49] None of these new theoretical endeavors, however, have altered his basic division between labor and inter-action and between instrumental and practical reason. They simply represent an elaboration of the conceptual foundations of the practical and emancipatory cognitive interests, that is, a continuation of certain themes developed in his earlier work. Moreover, as we shall see, Habermas has defended the fundamentally anthropocentric framework of his communication theory in his major reply to ecologically oriented criticisms of his theory of cognitive interests.[50] A critical discussion of these earlier ideas and categories (along with Habermas's response to his ecological critics) is therefore essential to understanding the anthropocentric framework of Critical Theory and assessing its relevance to ecocentric political theory.

THE ECOCENTRIC CRITIQUE

Although Habermas has made only occasional reference to the ecology crisis in his extensive writings, the general outlines of a Habermasian solution to the crisis are clearly discernible.[51] Although he has argued that, in a rational society, instrumental reason would be made subservient to the norms established by practical reason as a result of free discussion, he also insists that (whatever these norms) a rational society would continue to apply, *and would only apply,* instrumental reason to our dealings with the nonhuman world through human work and technology. Indeed, our environmental problems must necessarily be solved by the application of instrumental reason according to Habermas because that is the only kind of reason that Habermas considers to be efficacious in our dealings with nature from the point of view of our species' interest in survival.

The two most problematic aspects of Habermas's approach are (i) that it separates and privileges human emancipation vis-à-vis the emancipation of nonhuman nature; and (ii) that it claims that we can know nature (through science and technology) only insofar as we can control it. Indeed, Habermas's distinctions effectively serve to legitimate the continued exploitation of nonhuman nature, thereby endorsing rather than challenging dominant anthropocentric prejudices toward the nonhuman world.

According to Habermas's schema, a norm is considered "right" if it is achieved via a consensus reached between truthful, uncoerced, and rational human agents. It follows that if a "speech community" agrees, after free and rational discussion, to direct technology in such a way as to continue to manipulate and subjugate "external nature," then Critical Theory can raise no objection since its concept of emancipation has been exhausted (its exclusive concern being with human self-determination).

Now defenders of Habermas's theory, such as John Dryzek, have argued that a communicatively rationalized society would be more *conducive* to ecological rationality (i.e., defined by Dryzek to mean maintaining human life-support systems) than the piecemeal approach of liberal/pluralist democracies where political decisions concerning "who gets what, when, and how" are contested primarily by *interested* parties.[52] This is because a communicatively rationalized society provides for the proposal and rational acceptance of *generalizable* interests common to all humans. As Dryzek argues:

> the human life-support capacity of natural systems is *the* generalizable interest *par excellence,* standing as it does in logical antecedence to competing normative principles such as utility maximization or right protection.[53]

Dryzek adds the further qualifier that "the likelihood that ecological concerns will be reflected in social norms in communicatively rationalized settings could be enhanced, one suspects, if the community in question were small-scale and self-sufficient."[54]

This case is defensible in the terms in which it is presented, that is, from a human welfare ecology perspective. However, it is problematic from an ecocentric perspective because the moral referents in any consideration of ecological problems will only ever be the human participants in the dialogue. Indeed, Dryzek concedes the limitations of Habermas's Critical Theory in its endorsement of an anthropocentric and instrumental human orientation toward the *nonhuman* world:

> Habermas sees a discontinuity between the systems of the human world (potential subjects) and those of the natural world (inevitable objects). From the viewpoint of ecological rationality, this discontinuity is a misplaced decomposition of a non-reducible system.[55]

According to Habermas's schema, nonhuman entities—the objects of technical control—cannot be morally considerable subjects. His theory can therefore provide no basis for the preservation of species and ecosystems that serve no purpose for humans. As Whitebook summarizes it: "the proper norms for regulating the relations between society and nature would somehow follow from the communicatively conceived idea of the human good life without reference to nature as an end-in-itself."[56] Habermas has endorsed these comments as an accurate extrapolation of his theory.[57]

While Habermas acknowledges that many of us share an intuition of "sympathetic solidarity" with the nonhuman world, he is unable to work the interests of nonhumans into his theory in any meaningful way because it is *theoretically grounded* in human speech acts. He argues that the egalitarian reciprocity that he regards as implicit in *human* communication "cannot be carried over into the relation between humans and nature in any strict sense" because it presupposes that the referents are free and autonomous human subjects.[58] Habermas concedes that the range of communicative actions is broader than that of explicit human speech acts but argues that his approach enables us to grasp the *distinctive* features of human communication. But why ground communicative ethics in this limited way by focusing exclusively on this particular "differential imperative" (i.e., our communicative abilities)? As Anthony Giddens has argued, "the division we make between nature and culture is one that dissolves the intimacy with nature that is one of the richest forms of human experience."[59] Moreover, as Whitebook observes:

> The dignity and rights of the moral and legal subject have been secured by severing the subject from the realm of natural existence. Because they are characterized by self-consciousness or language, subjects are considered qualitatively different from the rest of natural existence. This is why they command respect and ought to be treated as ends-in-themselves. It is often feared that anything that threatens to disturb this distinction—*which the concept of nature as an end-in-itself certainly does*—also threatens the dignity of the subject.[60]

As we saw in chapters 2 and 3, there are many other plausible ways of grounding ethics that recognize the dignity of both human and nonhuman beings. We need to revise and extend Habermas's communication ethics to a full-blown ecocentric ethics that is informed not only by the internal related-ness and reciprocity embedded in human speech, but also by the internal relatedness and reciprocity embedded in ecological relations in general, which, in a very literal sense, sustain us all. The fact that the nonhuman world cannot participate in human speech should be no barrier to their spe-cial interests always being *considered* and respected by those who can partic-ipate in the dialogue. Indeed, as animal liberation theorists have pointed out, not all *humans* are able to participate in the rational speech community (e.g., the very young, the mentally ill, and the senile) although their interests are generally considered by those who do. It is not necessary to be a rational, speaking moral *agent* in order to be a morally considerable *subject,* as Habermas theory requires.[61]

Habermas's theoretical endeavors have been concerned to iron out dis-tortions in the Enlightenment project in order to *perfect* that project, namely, the pursuit of rational autonomy via the overcoming of all natural and social constraints on human thought and action. Ecocentric emancipatory theorists, in contrast, are concerned to revise this project in a quite different direction. The Enlightenment notion of rational autonomy—particularly the quest to overcome all natural constraints—is seen as fundamentally illusory since it denies the fact of humanity's embeddedness in nature. Accordingly, ecocen-tric theorists are concerned, among other things, to emphasize our continuity with and relatedness to the nonhuman world rather than our separation and differentiation from it, and to cultivate an orientation that recognizes that "the development and fulfilment of the part can only proceed from its com-plex interrelationship and unfolding within the larger whole."[62] Such an ori-entation should imbue all human activity—not only art, play, and contempla-tion, but also work, science, and technology. As Vincent Di Norcia argues, the split in Critical Theory can be resolved only by the development of "a more ecological, fraternal but still rational conception of the science and technics of nature."[63]

This brings us to the second problematic aspect of Habermas's theory—that we can know nature only in instrumental terms. This claim has attracted strong criticism from those who see his categories of thought as unduly limit-ed and/or as part of the cause of the environmental crisis rather than its solu-tion.[64] Some critics have rightly challenged the *objective* status of Haber-mas's technical interest and pointed to the historical specificity of the modern drive to control. As Henning Ottmann has pointed out, humans have not always assumed the Cartesian mantle of "masters and possessors of nature."[65] In any event, Ottmann argues:

> even if, in the name of the survival of the masses of contemporary
> humanity, we did not want to dispute the legitimacy of the modern
> type of mastery over nature entirely, nevertheless this does not mean

that we should accept *carte blanche* the will to control and its modern form. A will to control, whose legitimacy is based upon our need to survive and which is itself a threat to our survival, becomes dialectical. The technical interest in mastery over nature encounters a nature taking revenge upon the boundlessness of the will to control.[66]

A further reason why environmental problems are likely to remain intractable in a Habermasian society arises from Habermas's insistence that the technical cognitive interest in control must remain untrammeled by an alternative sensibility if it is to be successful in the terms in which it has been defined. Ecocentric theorists argue that the mere refinement of our ability to manipulate and control nonhuman nature will simply give rise to more "technological fix" solutions that will perpetuate or, at best, contain rather than solve, environmental problems since we can never be fully cognizant of all the interrelationships between the human and nonhuman worlds.[67] The Green revolution in agriculture, for example, once widely hailed as an example of how instrumental reason could alleviate world hunger, is now increasingly seen as creating ecological problems that ultimately have served to accentuate world hunger. Indeed, the ecology crisis may be seen as partly stemming from the extensive and overconfident application of instrumental reason to ecosystems. It this respect, it may be seen as a reflection of some of the inherent limitations of instrumental reason (particularly when applied on a grand scale) and of the need to cultivate an alternative human interest in nature. This new human interest in nature need not and ought not be circumscribed by the objectified image of nature that is called forth by instrumental reason.[68] As Di Norcia reminds us, "social and natural emancipation codetermine each other."[69]

Yet even if we were to assume that further technical refinements *would* succeed in protecting *human* welfare (after all, this technical interest in nature is rooted in the survival needs of the human species and our technical interventions in ecosystems would therefore have to be adjusted in response to threats to those survival needs), the consequences would be disastrous from the point of view of those species that are not presently or potentially useful to humankind. This is because Habermas's technical interest—just like his practical interest—leaves no room for the recognition of the intrinsic value of the nonhuman world. While Habermas accepts that an empathic orientation toward nonhuman nature might infuse art and recreation, he denies the need for a conceptual shift from anthropocentrism to ecocentrism in relation to our most significant dealings with nature in terms of impact on ecosystems (i.e., work, science, and technology).

To the above criticisms, Habermas has replied that he is simply making an epistemological—as distinct from an ethical—statement as to the type of reason that is capable of giving rise to theoretically fruitful knowledge, that is, knowledge that produces "efficacious results" from the perspective of our species' interest in material reproduction and survival.[70] We may, he argues, have noninstrumental encounters with nature (e.g., aesthetic experiences) but these encounters do not produce efficacious results in the way that instrumental reason does through its systematic observation, objectification, manipulation, and control of natural phenomena. Habermas agrees that "the moralization of our dealings with external nature" would indeed lead to the "reenchantment of the world" but disapproves of such a step on the ground that it would involve a regression, an undoing of the differentiation of knowledge that Habermas has categorized in his theory of cognitive interests as being the *progressive* outcome of the Enlightenment.[71]

Yet Habermas has not only failed to show that pure instrumental reason is always the most efficacious form of reason from the standpoint of human well-being and survival (as the Green revolution example demonstrates). He has also failed to show that instrumental reason is the *only* form of reason that can deliver theoretically fruitful knowledge of nature. As Alford has argued, medical anthropology is replete with examples of so-called primitive techniques of healing that are mediated by a "communicative attitude" toward nature and which are nonetheless efficacious in terms of achieving the intended results.[72] Similarly, the farming and fishing techniques of many traditional cultures are often more "efficacious" from a *long term* point of view than the modern agricultural, forestry, and fishing techniques that have so often replaced such traditional techniques. A more modern example can be found in the practice of biodynamic farming, which is mediated by a symbiotic and communicative relationship with the land. Although this practice cannot be fully explained by modern science, it is demonstrably efficacious from the point of view of results. Of course, these traditional and modern practices contain elements of instrumental reason; the point, however, is that they are also infused with other forms of reason that mediate and guide the technique and which, from the practitioners' point of view, are part and parcel of the technique's particular kind of efficacy.

These examples are not meant to imply that we must dispense with instrumental reason. Rather, they suggest that ecologically benign interventions in ecosystems are more likely to arise if instrumental reason is allowed to be thoroughly infused with and tempered by—rather than simply instructed by—normative considerations concerning human well-being and respect for other life-forms. More generally, they indicate that the dialectic between science and society is more complex and interwoven than Habermas's analytical categories can allow, notwithstanding his pioneering critique of scientism.

For example, Habermas is right to point to the existence of different human interests lying behind different forms of theoretical inquiry. However, he must be challenged in his insistence that only one very limited and anthropocentric kind of interest lies behind our scientific and technological endeavors. Indeed, it is quite common to find different kinds of interests determining different kinds of inquiry *within* the same branch of science. This is particularly evident within the science of ecology. Unlike Habermas, who would insist that ecologists can "know" their subjects (e.g., ecosystems, populations) only insofar as they can predict, manipulate, and control them, Donald Worster has shown that many ecologists approach their subject of study in the manner of a *partner in communication.* Science, Worster reminds us, is as divided as the rest of the Western civilization in terms of its orientation toward the nonhuman world.[73]

Alford has argued that Habermas's abstract treatment of the idea of science (shared also by Marcuse) is the legacy of Horkheimer and Adorno's *Dialectic of Enlightenment,* which saw science as "that fragment of reason concerned with human self-assertion."[74] Habermas continues this tradition by positing a direct and simplistic connection between science and technology, assuming that the role of pure science or "basic research" is always ultimately concerned to produce technically exploitable knowledge. Yet this is a very limited conception of science that ignores the role played by science in providing *meaning*—especially in shaping our understanding of our place in the cosmos. As Fox argues, modern science has both an instrumental aspect *and* a cosmological aspect. The latter provides us with "an account of creation that is the equal of any mythological, religious, or speculative philosophical account in terms of scale, grandeur, and richness of detail."[75] Moreover, we saw in chapter 3 that modern science has served to undermine anthropocentric assumptions by showing that humans are part of a seamless web of interrelationships, that there are no radical divides between the human and the nonhuman. We also saw how some of the most vigorous challenges to the notion of a detached scientific observer standing above and apart from the object of study have come from within science itself (e.g., in quantum mechanics and ecology).

At least Marcuse, for all his contradictory statements and vagueness concerning the issue of a new science (and despite his ultimate anthropocentrism), had a greater appreciation than Habermas of the historical relativity of human knowledge and the mutual interplay between different kinds of knowledge and different kinds of human interests. For example, just as science can shape the kinds of technologies we develop, it is also shaped by those very technologies (e.g., computers have enabled the development of chaos theory although they were not built for that purpose).[76] Similarly, just as science can be influenced by broad cultural paradigms, it can also help to change those very paradigms. As Di Norcia argues, "it is a half-truth to say that technologies are just utilitarian projections of bodily functions; they are also symbolic forms of self-expression and objects of self-inquiry."[77] A guided missile and a classical record embody vastly different forms of human self-expression.

It is surely not incongruous or regressive to suggest that a different and better science might result from a community of scientists who employ empirical-analytic modes of inquiry but who proceed on the basis of an ecocentric "interest" in the world. Such an interest would not only influence the types of problems and questions examined by such scientists, but also the way they go about their science, such as the types of theories they choose (given that theories are generally "underdetermined by the facts"), the types of "facts" they choose (given that sensory experience is underdetermined by sensory input), and the types of experiments and techniques they develop to test such theories.[78] As Evelyn Fox Keller has argued, scientists who have a "feeling for the organism," that is, approach their subject in a spirit of attentiveness, humility, and respect for the uniqueness of what is studied, can still produce reliable and sharable scientific knowledge about the natural order.[79] For ecocentric scientists, then, the test of "good" science would not be simply that it "works" in the sense of enabling humans to exploit the world around them more efficiently but rather that it "works" in the sense of enabling humans to live in ways that preserve and foster the health, safety, and well-being of both the human and nonhuman community. Far from being a mere "neutral" handmaiden of the polity, science itself might then become a further form of *resistance* to ecological degradation and the "colonization of the life-world."

THE "GOOD LIFE" REVISITED

The principal objection to Habermas's social and political theory has been that it is thoroughly human-centered in insisting "that the emancipation of human relations need not require or depend upon the emancipation of nature."[80] Although Habermas has moved beyond the pessimism and utopianism of the first generation of Critical Theorists and has provided firmer theoretical foundations for his practical and emancipatory cognitive interests, he has, as Whitebook points out, also "markedly altered the spirit of their project."[81] Yet it is precisely the "spirit" of the early Frankfurt School theorists (i.e., their critique of the dominant "imperialist" orientation toward the world, rather than their critique of a simplistically conceived idea of science) and their desire for the liberation of nature, that is most relevant to the ecocentric perspective. Despite Habermas's many theoretical innovations and departures from Marxism, ultimately he has strayed very little from the structure of the basic Marxist response to the environmental crisis presented in chapter 4 (whether orthodox or humanist). Such a response seeks to revolutionize social relations (while leaving intact our instrumental relationship to the nonhuman world) so that the forces of production can be rationally controlled by society as a whole.

Moreover, even within the human realm, Habermas is mainly preoccupied with the formal—as distinct from the substantive—ground rules for human emancipation. To be sure, modern Critical Theory clearly holds out the promise of cultural and ecological renewal by providing the space for the expansion of the moral and aesthetic spheres of communication vis-à-vis the technical sphere. (I say "promise" since Habermas, unlike Dryzek, pays very little attention to the types of institutions that might facilitate an ideal speech situation and the harmonious balancing of system-steering mechanisms and the life-world.[82]) Ecocentric theorists would agree that a rationalization of communication along Habermasian lines would indeed be conducive to protecting generalizable human interests (especially in smaller scale communities) and generating "cooperative" solutions to the environmental crisis that would avoid the familiar "tragedy of the commons" scenario. Such a rationalization would also be more likely to produce just solutions to the many social pathologies of modernity, ranging from crime and urban decay to poverty and unemployment. As Luke and White argue, Habermas's communicative ethics will enable the "deconstruction of managed meanings" by the corporate capitalist and bureaucratic state apparatus that "will help open the way for rethinking what autonomy in everyday life can mean for average producers and citizens in an informational age."[83]

But this communicatively rationalized social democracy is merely a necessary as distinct from a sufficient condition for emancipation writ large. Although one of Habermas's professed goals is to redeem the promise of the classical concept of politics by reviving the inquiry into the "good life" and restoring the art of *phronesis,* or practical reason, his approach has been essentially procedural rather than substantive. That is, he has failed to revive the classical tradition's pedagogical concern for the cultivation of a range of specific virtues in its citizenry (i.e., *in addition* to the civic virtue of democratic participation).[84]

Ecocentric emancipatory theorists, in contrast, have more in common with the classical tradition insofar as they are concerned to cultivate what might be called general "ecocentric virtues" (such as humility, compassion, knowledge of the local bioregion, and respect for the integrity and diversity of other life-forms) in addition to the civic virtue of participation. As we have seen in chapter 1, the ecological crisis has been identified not simply as a crisis of participation or survival but also as a crisis of *culture and character.* To these theorists, a radical reconception of our place in the rest of nature is not only essential for solving our planetary problems; it would also offer a surer path for human self-development. It is in this context that primary ecopolitical questions concerning human needs, technology, and lifestyles must be debated.

CHAPTER 6

Ecosocialism: The Post-Marxist Synthesis

INTRODUCTION

The late Raymond Williams once described the ecology movement as "the strongest organized hesitation before socialism."[1] Ecosocialism—a position Williams himself defended in his later writings—represents a concerted attempt to revise and reformulate the democratic socialist case in the light of the ecological challenge. Ecosocialists have also used this opportunity for theoretical stock-taking to respond to other significant challenges before socialism—challenges that form part of, but are not unique to, the ecological critique—in an effort to address the concerns of new social movements and recapture the "high ground" of emancipatory discourse. As Frieder Otto Wolf has put it, "a socialism without qualification will never again be able to become a hegemonic force within emancipatory mass movements."[2]

The three most significant challenges before socialism that ecosocialists have sought to address are (i) the historical legacies of bureaucratization, centralization, and authoritarianism; (ii) the problematic role of the working class as the agent of revolutionary change; and (iii) disillusionment with the traditional "productivist" and growth oriented socialist response to the indignities of poverty, which has usually been to augment the economic power of the State, seek a better mastery of nature through modern scientific techniques, and step up production. Ecosocialists have sought to respond to these historical legacies by reasserting the principles of democratic self-management and production for human need. According to Williams, "this is now our crisis: that we have to find ways of self-managing not just a single enterprise or community but a society."[3]

The ecosocialist theory presented in this chapter represents the most influential family of socialist thought in Green circles. It has emerged from a critical dialogue between Marxist orthodoxy, various currents of Western Marxism, and Western social democracy, on the one hand, and the radical environmental movement, on the other hand. The resulting body of theory may be described as largely post-Marxist insofar as it is highly critical of orthodox Marxism (and much Western Marxism) but is not *anti*-Marxist. That is, many theorists within

119

this tradition occasionally draw on Western Marxist insights alongside other older traditions and contemporary strands of socialist thought, including utopian socialism, the self-management ideas of the New Left, and socialist feminism. While ecosocialists also share the anti-capitalist and self-management orientation of ecoanarchists, they generally argue (contra ecoanarchists) that the State must play a key role in facilitating the shift toward a more egalitarian, conserver society.

The growing influence of ecosocialist ideas within the Green movement (most notably in Europe and Australia rather than in North America) has rendered the popular Green slogan "neither left nor right" somewhat problematic.[4] While this slogan originally served to publicize the Green movement's efforts to find a distinct, third alternative to the growth consensus of capitalism and communism, it has since served to generate a lively and sometimes acrimonious debate within the Green movement concerning the proper political characterization of Green politics. Ecosocialists argue that "elements of the Left are the natural allies of the Greens" and that only a new ecosocialism can provide a feasible, third alternative to existing capitalism and communism.[5] In particular, ecosocialists have mounted a challenge to the presumed left-right ideological neutrality of Green politics by pointing out the various egalitarian and redistributional (and hence "leftist") measures that are needed to ensure an equitable transition toward a conserver society. Indeed, many such measures—such as the redistribution of resources from developed to developing countries, the sharing of work, and the implementation of a guaranteed minimum income scheme—are already included in many Green party platforms.[6] With respect to these kinds of issues, Green political aspirations can indeed be fairly described as "more left than right."

However, as we saw in part 1, to approach Green politics only through the prism of the conventional left/right ideological cleavage is to miss the most distinctive critical edge of Green thought, namely, the critique of the cornucopian and anthropocentric assumptions of modern political thought. Ecosocialism, as we shall see, has challenged the former but has made no substantive inroads into the latter. This notwithstanding, the possibility of some theoretical bridge building between ecosocialism and ecocentrism remains open at the level of both ecophilosophical orientation and socioeconomic critique. This does not, however, necessarily extend to the ecosocialist prescription for change. In particular, there are many Green theorists who accept the ecosocialist critique of capitalism but who do not accept that this critique necessarily requires the curtailment of the market economy to the degree envisaged by many ecosocialists.

THE ECOSOCIALIST CRITIQUE

There are many points of convergence between the democratic socialist critique of private and state capitalism and the radical ecology movement's critique of industrialism that point toward the possibility of a theoretical synthesis of socialism and ecology.[7] Indeed, it is this convergence that has prompted the development of ecosocialist theory. Ecosocialists argue that it is the competitive and expansionary dynamics of the capitalist system that is largely responsible for the ecology crisis. However, they are critical of nonsocialist Greens for neglecting class politics and failing to develop "an analysis of power" in their "new ecological paradigm." Ecosocialists consider that such an analysis is essential if a fundamental opposition to the present means of production, distribution, and exchange is to be mounted.[8] For example, Joe Weston has argued that it is the accumulation of wealth and its concentration into fewer and fewer hands that is the main cause of both poverty *and* ecological degradation.[9] He goes on to insist that "it is time that greens accepted that it is capitalism rather than industrialism *per se* which is at the heart of the problems they address" (Weston, like most ecosocialists, regards Soviet Russia as having practiced "state capitalism" rather than socialism).[10] Similarly, David Pepper is critical of what he calls "new paradigm" Greens for focusing on ecologically degrading methods of production rather than on who owns and shapes such methods and the social relations that stem from them.[11] Indeed, most ecosocialists regard the ecology crisis as but one, albeit increasingly significant, item in a much broader agenda. As Gorz puts it: *"the ecological movement is not an end in itself, but a larger stage in the larger struggle* [i.e., to overcome capitalism]."[12]

It is undoubtedly the case that the expansionary dynamics of capital accumulation have led to widespread ecological degradation and social hardship. One of the most basic reasons for this is that the profit motive demands that firms "grow or die." This imperative for continual economic growth does not respect physical limits to growth or ecological carrying capacity. The upshot is that there are many situations in which market rationality gives rise to "negative externalities" such as resource depletion and pollution, which are the unintended and unwanted side-effects of capital accumulation. These externalities are usually borne by those who do not produce or consume the goods or services in question. However, there are some situations, such as the exploitation of common property or "free environmental goods," where market rationality creates outcomes that are worse for *all* agents. This is illustrated in game theory by the Prisoners' Dilemma and by Garrett Hardin's oft-quoted parable of the "tragedy of the commons."[13]

Capitalism also generates uneven development (both within and between nations). Capital is now highly mobile, and investment is governed by absolute profitability rather than national affiliation or a concern to develop mutually advantageous trading arrangements. This increasing international mobility of capital has led to serious trade imbalances and external indebtedness on the part of Third World nations, which has generated widespread ecological degradation and poverty.

Finally, market rationality gives priority to short-term interests over long term interests through the practice of "discounting the future." This effectively creates a structural bias *against* future generations. Indeed, the goal of profit maximization encourages the liquidation or depletion of both renewable and nonrenewable resources and the movement of the capital thereby gained into new ventures rather than the sustainable or prudent harvest of resources over time.[14]

Ecosocialists argue that the logic of capital accumulation is fundamentally incompatible with ecological sustainability and social justice. Accordingly, they argue that capitalism must be largely replaced with a nonmarket allocative system that ensures ecologically benign production for genuine human need. The real challenge facing ecosocialists, however, is how to develop new, democratic, and noncentralist social institutions that are able to give expression to ecosocialist values such as self-management, producer democracy, and the protection of civil and political liberties. These issues will be explored later in this chapter. For the moment, it will be helpful to to draw together the lessons ecosocialists have learned from the failures of existing communist regimes and the revisions they have made to socialist theory in the light of the ecological challenge.

Farewell to Scientific Socialism and the Economic Growth Consensus

Ecosocialists accept that there are both ecological and social limits to growth and accordingly they reject the economic growth consensus of conservative, liberal, social democratic/labor, and communist parties. According to Williams, the central ecological problem created by market capitalism and the economies of existing communist regimes is that there is "an effective infinity of expansion in a physically finite world."[15] In addition to their rejection of the indiscriminate commitment to mass manufacture and increased consumption, ecosocialists also share the early Frankfurt School's rejection of "scientific socialism" as being unduly optimistic in believing in the unlimited power of scientific understanding, technical control, and the mastery of nature. In this respect, ecosocialists wish to avoid replacing the compulsion of the market with bureaucratic domination on the ground that both capitalist and communist economies dominate both people and nonhuman nature. They are generally critical of large-scale institutions and alienating, "inappropriate," or destructive technologies and advocate what Ryle has called "eco-contraction" or ecological restructuring (in Ryle's case, this entails, inter alia, the gradual dismantling of major ecologically destructive industries such as the automobile, chemical, and defence industries).[16]

Unlike orthodox eco-Marxists, ecosocialists question both the capitalist relations of production *and* the capitalist forces of production. In particular, they argue that the orthodox Marxist prophecy that conventional socialist strategy had counted on—that the intensification of the contradictions between the capitalist forces and relations of production will finally be resolved by the industrial proletariat taking over the forces of production— will not end the ecological crisis or human alienation. This is because the expanded forces of production are ecologically degrading and, in any event, do not lend themselves to collective appropriation insofar as "there can never be effective self-management of a big factory, an industrial combine or a bureaucratic department. It will always be defeated by the rigidity of techni- cal constraints."[17] In short, ecosocialists maintain that the mere appropriation of the capitalist forces of production would result in a new ruling class taking over the machinery of domination.

In accepting the early Frankfurt School's critique of instrumental reason, ecosocialism reveals, in varying degrees, a *partial* return to a pre- Marxist/romantic critique of industrialization. Whereas human interactions with the natural world under Marxism were always as Producer, Williams declares that under ecosocialism our interventions must now proceed

> from a broader sense of human need and a closer sense of the physi- cal world. The old orientation of raw material for production is rejected, and in its place there is the new orientation of livelihood: of practical, self-managing, self-renewing societies, in which people care first for each other, in a living world.[18]

This kind of general reorientation away from instrumental reason is funda- mental according to Williams, "for it is the ways in which human beings have been seen as raw material, for schemes of profit or power, that have most radically to be changed."[19]

The Problematic Role of the Working Class

Ecosocialist theorists recognize that the industrial working class has not only shrunk in size relative to other classes but has also become increasingly con- servative by virtue of its economic dependency on the capitalist order. Indeed, most ecosocialists accept that the working class—whatever its histo- ry—is no longer the central agent of progressive social, cultural, and political change, and they concede that such change is more likely to emanate from a broad front of allied new social movements that operate outside the tradition- al labor movement and that are not easily defined by their class location. André Gorz, in particular, has argued that the industrial proletariat cannot become the revolutionary force heralded by Marx since it has turned into a mere replica of capital, exercising functional but not personal power. Indeed, in *Farewell to the Working Class,* Gorz anticipates that the traditional, skilled proletariat will become more disciplined, conservative, and privileged over time as increased automation reduces the number of working-class jobs. As a result there will be a swelling in the ranks of what Gorz has called the "nonclass" or "post-industrial neo-proletariat," a "class" that encompasses all those who have been expelled from manual and intellectual work as a result of automation and computerization as well as those who are marginally employed and who have no real class identity or job security.[20] Gorz extends this line of argument in *Paths to Paradise* where he maintains that increasing automation and the microelectronic revolution are reducing the quantity of labor required for most material production and breaking down the direct contact between worker and matter. Full employment has become an unrealizable goal, yet Gorz argues that it is continually pursued as an ideological tool to "maintain the relations of domination based on the work ethic."[21] This leads inevitably to an increasing split in the active population between

> on the one hand, acting as a repository of industrialism's traditional values, an elite of permanent secure, full-time workers, attached to their work and their social status; on the other, a mass of unemployed and precarious casual workers, without qualifications or status, performing menial tasks.[22]

This latter "nonclass" occupies a pivotal place in Gorz's analysis in that it is seen as the prefiguration of a different kind of convivial community—beyond economic rationality and external constraint—that constitutes a potentially emancipatory extension of an already developing process.[23] Indeed, Gorz's ultimate project is to abolish wage labor on the ground that there is no dignity to be had in the modern wage labor relationship. Instead he argues that true dignity and self-determination can only be found in autonomous spheres of production (i.e., in the neighborhood rather than in the factory).

However, unlike Gorz, most ecosocialists see the potential for social change emanating from a much broader alliance of new social movements *working together with the labor movement.*[24] While ecosocialists concede that the labor movement's "productivist ideology" has traditionally not recognized the experiences of other disadvantaged groups and classes (e.g., welfare recipients, women, and ethnic minorities) or environmental problems beyond the workplace, they nonetheless insist that effective and lasting change will not come about without the support of the union movement, indeed, the majority of ordinary "working people."[25] In short, ecosocialists point out that although environmental problems go beyond class issues, they still contain a class dimension, which cannot be glossed over.[26]

However, building such an alliance between the New Middle Class radicals that support new social movements, on the one hand, and the working class, on the other hand, is no easy task. As Williams observes, the predominantly middle-class membership of new social movements must confront the fact that the "effective majority" will remain committed to the dominant system *so long as they have no practical alternative.*[27] Williams (unlike some ecosocialists), however, dismisses as absurd the claim that new social movements are elitist or that their claims are in conflict with the interests of the working class. The reason why the demands of the New Middle Class differ from those of the working class is primarily a matter of different social experience and different access to information. As Williams explains,

> the fact that many of the most important elements of the new movements and campaigns are radically dependent on access to independent information, typically though not exclusively through higher education,...[means] that some of the most decisive facts cannot be generated from immediate experience but only from conscious analysis.[28]

In this respect, Williams rightly points out that unless the Green movement can generate "serious and detailed alternatives at these everyday points where a central consciousness is generated" (i.e., the local, practical, and immediate interactions of the "effective majority" of working people), then the issues raised by the Green movement will remain marginalized.[29] Williams argues that an important task for ecosocialism, then, is a "critical engagement" with the labor movement in order to prepare the way for a broad Green/labor alliance that will represent the "general interest" as distinct from the interests of a particular class. Ecosocialists divide, however, on the question as to whether to pursue this critical engagement through the established social democratic and labor parties, through the fledgling Green parties and the Green movement, or through a new grassroots rainbow movement.

As part of the move to widen the narrow, "productivist" focus of the traditional labor movement, ecosocialists have also sought to expand the traditional democratic socialist preoccupation with class and producer democracy to include cultural renewal and the revitalization of civil society. This entails the promotion of new attitudes to work (aided by job-sharing and a reduction in the length of the working week), health, lifestyle, and sexuality. However, unlike many ecoanarchists and ecofeminists, most ecosocialists generally avoid any discussion of the need to develop a "new ecological paradigm" (ecocentric or otherwise), much less new forms of Western spirituality, and tend to adopt instead a secular approach that emphasizes the cultivation of public virtue or good citizenship rather than "inner awakening."

The New Internationalism

In responding to the ecology crisis, ecosocialists have sought to explore a broader range of contradictions than those based simply on class. For example, most share Rudolf Bahro's analysis that the "external" contradiction between humanity and the rest of nature and between "North" and "South" have become more pressing than the "internal" contradictions between capital and labor within the developed countries of the First World.[30] The resolution of these contradictions is seen to require a "new internationalism" that accepts that we cannot use the standard of living attained by the average family in the First World as a model to be pursued for all of humanity since this would put an intolerable ecological strain on the planet. Ecosocialists therefore argue that the transition toward a conserver society must begin in the "affluent society."[31] According to Williams, the deepest changes must come from the First World not only in the form of conservation and the production of more durable commodities "but also in their deep assumption that the rest of the world is an effectively vacant lot from which they extract raw materials."[32] Accordingly, ecosocialists argue for the redistribution of wealth not only within nations but also internationally between the developed and developing countries in order that all peoples of the world may pursue a lifestyle that is within the Earth's carrying capacity.[33] A cornerstone of this "new internationalism" is a redefinition of human needs that is global in scope. According to Ryle, ultimately, human needs have to be defined at a level that enables both present and future generations of humans to enjoy an equivalent measure of health and autonomy.[34] Ryle suggests that a priority in this exercise should be the establishment of an agreed set of basic needs (i.e., education, health care, energy, basic infrastructural requirements such as sewage and water supply, housing, and transport), so that steps can then be taken to ensure that everyone has these basic needs met in both rich and poor countries.[35] Unlike democratic socialists, however, *eco*socialists address social and economic deprivation by means other than expanding production. That is, they seek to meet human needs in ecologically benign and sustainable ways in order to bring overall resource consumption down to a level that is compatible with global justice and ecological integrity. This, of course, is a much more challenging task than simply stepping up production and providing *more* commodities and social welfare services.

Ecosocialists argue that Third World solidarity can be achieved by promoting greater self-reliance in both the North and South. This strategy requires partially delinking the economies of the North and South by reducing the volume of international trade, disarming, and increasing aid to developing countries. However, as Frankel and Ryle note, ecosocialists must also encourage international cooperation to ensure control of transnational corporations and financial institutions, which will require parallel and reciprocal moves by other nations if it is to be effective.[36]

The major thematic innovations of ecosocialism—the rejection of the economic growth consensus, the emphasis on ecologically benign production for human need, the attempt to widen the productivist outlook of the labor movement and encourage a critical dialogue between the labor movement and new social movements, and the new internationalism—together represent a major overhaul of socialist thought. Moreover, these theoretical revisions place ecosocialism squarely within the spectrum of Green or emancipatory political thought.

However, as we shall see, ecosocialism has self-consciously declined to step across the anthropocentric divide and embrace an ecocentric perspective. Instead, most ecosocialists have rejected the need for a "new ecological paradigm" and have argued that socialist thought provides a sufficient repository of values for ecological and social reconstruction.

THE MEANING AND LESSON OF ECOLOGY ACCORDING TO ECOSOCIALISM

The heart of the philosophical difference between ecocentrism and ecosocialism concerns the meaning and relevance of ecology to emancipatory theory. Ecosocialists regard the demands of the environmental movement for a safe and healthy environment as but a subset of the modern radical project. In particular, we have seen that many ecosocialists regard the radical environmental movement (by this they mostly have in mind what I have characterized in chapter 2 as the human welfare ecology stream) as part of a larger struggle to *overcome capitalism*. The radical environmental movement is seen to be part of that larger struggle because it highlights the incompatibility of market rationality with ecological limits by revealing the many ways in which the economic "externalities" of private capital have seriously compromised human welfare, health, and survival.[37]

What is at stake, and what is now attainable, according to ecosocialism, is the full realization of *human autonomy* within a safe and healthy physical environment and a democratic and cooperative social environment. Significantly, most ecosocialists reject the idea that ecology can effect a fundamental paradigm shift in political theory along the lines suggested by many Green theorists.[38] Ecosocialists argue that the preoccupation with ecological principles leads to an excessive preoccupation with "nature protection" and deflects attentions away from the *social* origins of environmental degradation.[39] What must be grasped, they argue, is that the "environment" is an essentially *human* context that is *socially* determined rather than something before which we must humbly "submit."[40] More generally, ecosocialists argue that if we wish to retain a commitment to the modern political ideals of justice, equality, and liberty, then we must look to the lessons of human history (as interpreted by the various tributaries of socialist thought) rather than natural history. Indeed, Gorz has gone so far as to declare that it is "impossible to derive an ethic from ecology."[41]

In support of the argument that ecological principles cannot provide the basis for a new politics, ecosocialist theorists frequently point out that it is possible to have a society that respects ecological limits but is undemocratic and authoritarian.[42] As we have seen, ecosocialists reject the claimed "newness" of the Green movement and the idea that it has transcended old political rivalries and instead point out the continuities between Green politics and many strands of socialism.[43] The only "newness" of the Green movement is seen to reside in its recognition of "ecological limits"—something that ecosocialists agree cannot be ignored.[44] "The point," argues Gorz, "is not to deify nature or to 'go back' to it, but to take account of a simple fact: human activity finds in the natural world its external limits."[45]

Yet the ecosocialist argument that it is impossible to "derive" an ethic from ecology is misleading and creates an overdrawn opposition: that eco-centrism represents a naive form of authoritarian ecological determinism, while ecosocialism recognizes the active presence of humankind in constructing and shaping the environment. The environmental ethics of ecosocialism and ecocentrism are both *informed* by ecological insights, but the environmental ethic of ecosocialism is simply a different and more limited kind of environmental ethic than that of ecocentrism. That is, the ecosocialist ethic is a prudential ethic that largely represents an amalgamation of the resource conservation and human welfare ecology perspectives explored in chapter 2, both of which are informed by the life sciences such as ecology but which ultimately rest on anthropocentric *norms* of human autonomy, health, and welfare. Ecocentrism is also informed by the life sciences, but it, too, finds its ultimate justification in a normative rather than scientific framework. (As we saw in chapter 3, to appeal to nature as known by the science of ecology rather than to ethics as the ultimate arbiter of a Green political theory is misguided and does not in itself amount to a justification for a particular political posture.) This ecocentric normative framework subsumes the human-centered ecosocialists' norms of autonomy, health, and welfare in a broader ecological framework that seeks the mutual flourishing of *all* life-forms. Such a perspective does not seek to downgrade human creativity nor deny the extent to which humans influence ecological and evolutionary processes. Rather, it asks that we employ our creativity to develop technologies and lifestyles that allow for the continuation of a rich and diverse human *and* nonhuman world. Ecosocialism, in contrast, may be seen as merely fusing human welfare ecology with democratic socialism, but transcending neither.

To return to the ecosocialist critique, if the only "lesson" provided by ecology is one of physical limits to growth, then it is indeed possible to have a range of different political regimes—including authoritarian and fascist ones—that observe such limits.[46] Robert Heilbroner's *An Inquiry into the Human Prospect,* which is essentially concerned with human survival, is a clear case in point.[47] Of course, an authoritarian regime might be successful—at least in the short term—in ensuring human survival (or, more likely, the survival of certain privileged classes of humans) and quite possibly the (indirect) survival of many nonhuman life-forms. However, it would achieve this by severely restricting opportunities for democratic participation and self-determination—a route that is incompatible with the general ecocentric norm of mutual unfolding of *both* the human and nonhuman worlds. For ecosocialists to reject ecocentrism on the ground that it does not rule out fascism is to miss the *inclusive* nature of the ecocentric norm of "emancipation writ large."

The ecosocialist rejection of the idea of a "paradigm shift" in Green political theory is generally correct insofar as it applies to inter-human struggles. When viewed from the perspective of the traditional political spectrum, ecocentrism is, and must continue to be, generally "more left than right" in contending with old political rivalries based on differentials in wealth, power, and social privilege. However, ecocentric theory is most certainly new in the way it seeks to reorient humanity's relationship to the rest of nature. In this respect, it represents a new constellation of ideas that challenges the anthropocentric assumptions of post-Enlightenment political thought and calls for a more radical reassessment of human needs, technologies, and lifestyles than ecosocialism.

To be sure, we have seen that ecosocialism has itself travelled some distance down this new path insofar as it has acknowledged the many ways in which capitalism objectifies and commodifies both people *and* nonhuman nature.[48] However, the ecosocialist critique of instrumental reason, like that of the Frankfurt School, ultimately comes to rest on the human-centered argument that it is wrong to dominate nature, because it gives rise to the domination of people. For example, Gorz, in noting that the disregard of ecological limits will often set off an unwelcome ecological backlash, argues that

> it is better to leave nature to work itself out than to seek to correct it at the cost of a growing submission of individuals to institutions, to the domination of others. For the ecologist's objection to system engineering is not that it violates nature (which is not sacred), but that it substitutes new forms of domination for existing natural processes.[49]

Of course, this ecosocialist concern for human autonomy is laudable in and of itself. From an ecocentric perspective, however, it means that the case for the recognition and protection of nonhuman species is activated only when it can be shown to facilitate human emancipation. As Rodman has observed, Gorz has an intuition that we should "let nature be" not because it is sacred or has its own relative autonomy but because its makes *us* freer.[50] While such an argument has a place within ecocentric theory (indeed, it serves to bolster ecocentric theory in that it shows that the flourishing of human *and* nonhuman life need not be a zero-sum game), it provides no defence for threatened nonhuman species in those cases where there is no appreciable link with human domination and where such species appear to provide no present or potential use or interest to humankind. Moreover, as we saw in our critique of the human welfare ecology perspective in chapter 2, such an argument also serves to reinforce anthropocentric attitudes. As John Livingston has aptly put it, at best, wildlife might "emerge as a second-generation beneficiary" from human welfare ecology reforms.[51] In this respect, Raymond Williams's views on wildlife preservation are telling:

> we are not going to be the people…who simply say "keep this piece clear, keep this threatened species alive, at all costs." The case of a threatened species is a good general illustration. You can have a kind of animal which is damaging to local cultivation, and then you have the sort of problem that occurs again and again in environmental issues. You will get the eminences of the world flying in and saying: "you must save this beautiful wild creature." That it may kill the occasional villager, that it tramples their crops, is unfortunate. But it is a beautiful creature and it must be saved. Such people are the friends of nobody, and to think that they are allies in the ecological movement is an extraordinary delusion.[52]

This, of course, is consistent with my identification of wilderness or wildlife preservation as one of the "litmus tests" that enables us to distinguish ecocentric from anthropocentric Green theorists. That is, wherever there is an apparent conflict between human interests and the interests of nonhuman species (in this case the protection of wildlife) that appear to be of no use to humankind, ecosocialists *consistently dismiss* nonhuman interests.

Similarly, to the extent that ecosocialists have contributed to the human population debate (the other "litmus test" issue), it is usually by way of a critique of what are seen as the "neo-Malthusian" arguments of population control advocates such as Paul Ehrlich—a critique that follows the spirit, if not the letter, of Marx's critique of Malthus.[53] According to this argument, the real causes of resource scarcity, famine, and environmental degradation are not the existence of too many people or the limited carrying capacity of the Earth but rather social factors such as the maldistribution of resources and inappropriate technology, which arise under the capitalist mode of production.[54] The ecosocialist solution, then, is not population control but the replacement of capitalism with a cooperative social order that uses ecologically appropriate technologies for the satisfaction of human need.[55]

From an ecocentric perspective, the ecosocialist response goes only part of the way toward addressing the population problem. First, it fails to consider the many ways in which growing absolute numbers of humans can magnify environmental degradation and therefore impair the overall quality of human life. Second, it fails to consider the impact of growing absolute numbers of humans on the nonhuman community—a limitation that arises from the exclusive ecosocialist preoccupation with human welfare. The environmental impact of humans is a function not only of technology and affluence (i.e., level of consumption), but also absolute numbers of humans.[56] From an ecocentric perspective, it is not enough simply to wait for the "demographic transition" (i.e., the lower birth and death rates that usually follow improved living standards) to achieve a stable and well-fed human population, because the price of such a transition is further widespread ecological degradation and species extinction. To minimize ecological degradation during this transition period, ecocentric theorists argue that it is necessary to bring about, *in addition to* technological and distributional reforms and a lowering of resource consumption, a wide range of humane family planning measures with a view to stabilizing and then reducing human population. "Humane family planning measures" include free contraceptives and free birth control information and counseling; affirmative action to improve the status and social opportunities of women; and ecological education campaigns that explain, inter alia, the impact of human population growth on ecosystems and the need to reduce the size of families to one or two children. The synergetic effect of introducing ecologically benign technologies and lowering energy and resource consumption *as well as* lowering the birth rate would have a much more dramatic result in terms of lessening environmental degradation and protecting biotic diversity than would the more limited ecosocialist solution.

The foregoing critique of the anthropocentric premises of ecosocialist thought does not require a rejection of either the entirely defensible socialist concern to find an allocative system that ensures production for genuine human need *or* the more general and equally defensible concern to seek the mutual self-realization of all humans. Quite the contrary, both of these concerns fall naturally into the orbit of the ecocentric perspective defended in this inquiry. Indeed, there is already a strong resonance between ecocentric social goals and key ecosocialist goals such as the new internationalism, democratic participation, and ecologically sustainable production for human need. Moreover, both ecocentrism and ecosocialism reject an atomistic model of reality in favor of a reciprocal model of internal relations (albeit with different ethical horizons and implications).[57] These resonances open up the possibility of theoretical bridge building between ecocentrism and ecosocialism. That is, many ecosocialist principles and arguments can be selectively incorporated into the broader theoretical framework of ecocentrism *once they are divested of their anthropocentric limitations.* (This possibility would not, however, extend to the more aggressive anthropocentric categories of Marxist thought that have been rejected in chapter 4, nor to the overly rigid theoretical categories of Habermas that have been rejected in chapter 5.)

The upshot of such theoretical bridge building for ecosocialism would be a widening of its field of moral considerability so that it reaches beyond the human community to include all of the myriad life-forms in the biotic community. In particular, this would mean a broadening of the ecosocialist approach to wilderness protection and human population growth in accordance with ecocentric goals. More generally, it would mean a broadening of the context of political, economic, and technological decision making so that human interests are pursued, wherever practicable, in ways that *also* enable other life-forms to flourish.

The upshot for ecocentrism would be a strengthening and broadening of its political and economic analysis that would make it better equipped to determine the kinds of institutional changes and redistributive measures that would be required to ensure an equitable transition toward a sustainable and more cooperative society. It would also enable ecocentrism to anticipate and address in a more concerted way the various forms of opposition that are likely to be encountered in the attempt to give practical expression to ecocentric emancipatory goals. In this respect, ecosocialists are right to argue that capital will not be placed at the service of emancipatory goals without increasing government intervention in the market and without a gradual democratization of the economy. As we shall see in the following chapter, the ecosocialist discussion of the potential "enabling role" of the State in facilitating the realization of emancipatory goals provides an informative and pragmatic counterpoint to the anti-statism and excessive idealism of many ecoanarchist theorists, notwithstanding the stronger ecocentric orientation of the latter.

THE ECOSOCIALIST AGENDA

Having identified the areas of concern shared by ecosocialism and ecocentrism, it now remains to explore the challenging problem of how to provide feasible alternatives to modern capitalism and communism. As we have seen, ecosocialists emphasize the need to develop long term socioeconomic solutions that will bring the economy under more democratic control to enable ecologically sustainable production for human need. Williams, for example, looks forward to a redefinition of socialism that entails "a positive redemption of the central socialist idea of production for equitable use rather than for either profit or power."[58] This entails "the long and difficult move away from the market economy"; a shift in "production towards new governing standards of durability, quality and economy in the use of non-renewable resources"; and "as a condition of either of the former, we have to move towards new kinds of monetary institutions, placing capital at the service of these new ends."[59]

However, there is no unanimity among ecosocialists on the matter of detailed alternatives. All that can be safely generalized is that ecosocialists argue that the fulfilment of basic needs and the provision of social services should be in some way funded by the wealth produced by society as a whole. Although many ecosocialists acknowledge that a market economy has certain advantages over a planned economy in terms of efficiency and flexibility in the satisfaction of consumer wants, they argue that these advantages are overshadowed by the serious contradictions between market logic and ecological imperatives. As Ryle explains:

While market-like mechanisms might continue to play an important role—providing consumer choice and flexibility in the supply of commodities—in an ecologically planned economy, these central economic functions would need to be planned in [a] directly political fashion.[60]

Indeed, most ecosocialists tend to argue that if we are to avoid both the "tragedy of the commons" and extreme wealth differentials, then we need to move toward a planned economy, provided such planning is of a kind that provides for full community participation. To this end, most ecosocialists advocate a combination of state and local community economic planning; democratically controlled public enterprises; state regulation of the financial sector; self-managing worker cooperatives; and an informal "convivial sector." Some ecosocialists also support a small business sector, although profit accumulation and size are to be closely controlled.

Ecosocialists regard the State as playing a vital role in controlling the operation of market forces and in laying down the framework for a socially just and ecologically sustainable society. According to Ryle:

If one is honest,...about the objectives which an ecologically enlightened society would set for itself, it is difficult to avoid concluding that the state, as the agent of the collective will, would have to take an active law-making and -enforcing role in imposing a range of environmental and resource constraints.[61]

This entails giving up, in the name of the "common good," a range of Western freedoms concerning the use of private capital as part of the process of ecological restructuring. As Ryle argues:

Above all, it calls into being a collective subject, a "we," able to make political and cultural decisions directly, and this implies the transcendence of the atomised individualism of the marketplace as ultimate arbiter.[62]

As we shall see in the following chapter, this aspect of the ecosocialist case stands in stark contrast to the strongly anti-statist position of ecoanarchists. That is, ecoanarchists would argue that the collective "we" is the local community rather than the State and that society is best transformed through popular struggles, exemplary action, and local self-help initiatives. Indeed, many ecoanarchists wish to see the abolition (rather than just the shrinking) of the modern nation State on the grounds that it is inherently hierarchical in usurping the decision-making power of the local community.

Although ecosocialists support the goal of community empowerment, they argue that this goal would be facilitated rather than thwarted by the State by means of protecting civil liberties, providing long-term economic planning, redistributing resources between classes and regions, and providing international diplomacy. Indeed, Frankel has argued that "democracy would not survive the abolition of state institutions."[63] Moreover, ecosocialists argue that some degree of bureaucratic administration is inevitable if economic and ecological planning is to proceed. They argue further that the potential for bureaucratic domination or political abuse in an expanded State would be offset by parallel moves that extend the opportunity for democratic participation in all tiers of government.

For example, the ecosocialist discussion paper *New Economic Directions for Australia* envisages that the increased planning powers of the State would be counterbalanced by "a radical restructuring of administrative and political institutions to minimise bureaucratic structures and processes and maximise public participation in planning and decision-making at all levels."[64] This would be facilitated by the establishment of consumer and citizen initiative councils at the state level and a federal watchdog commission to investigate public administration.[65] The responsibility for directing and coordinating all public instrumentalities engaged in productive activity would be assumed by a National Enterprise Commission, which would be representative, publicly accountable, and subject to public review every three years.[66]

André Gorz has defended a more controversial "dual economy" based on a fusion of the ideas of the young Marx and Ivan Illich. This dual economy is to be constituted by (i) the "sphere of heteronomy," which deals with the social production of necessities and the material reproduction of society and corresponds to the "realm of necessity" (to be planned and managed by the State); and (ii) the "sphere of autonomy," which provides the space for creative and convivial activity and corresponds to the "realm of freedom" (and civil society). The purpose of the heteronomous sphere—socially administered production—is to secure, and where possible enlarge, the autonomous sphere—local convivial activity.[67] Accordingly, Gorz argues that the heteronomous sphere would need to take maximum advantage of automation, specialization, division of labor, economies of scale, and computerization in order to minimize socially necessary labor and maximize autonomous activity (where economic logic need not apply). Gorz envisages that socially necessary labor would be ecologically benign and would produce only necessary, durable commodities. Everyone would be required to perform necessary labor (which would provide the source of a guaranteed income for all), but such labor would be reduced to a minimum and accordingly would no longer be the center of everyone's life. The guaranteed income generated by socially necessary labor would represent an equitable distribution of the wealth created by society's productive forces considered as a whole—which individuals have combined to produce through their shared, intermittent work.

According to Gorz, both Illich and Marx envisaged a reduction of working time and hence "the utmost expansion of the sphere of autonomy."[68]

However, whereas Marx foresaw the withering away of the State under communism, Gorz envisages the continuation of the State as the centerpiece of a post-industrial political economy that will make possible the flourishing of an ecologically benign, "convivial society."

A crucial distinction in Gorz's dual economy is that between the systematic and collective needs of society and the ethical norms of individuals and small communities.[69] On the basis of this distinction, he envisages that the sphere of heteronomy would be planned by the State and governed by technical imperatives; these imperatives are regarded as simply the function of "external necessity" rather than as ethical norms of a kind chosen by self-determining individuals. As Gorz explains:

> In the same way that economics is concerned with the external constraints that *individual* activities give rise to when they generate unwanted *collective* results, ecology is concerned with the external constraints which economic activity gives rise to when it produces environmental alterations which upset the calculation of costs and benefits.[70]

The upshot is that both economics and ecology are seen as scientific tools that measure different levels of efficiency; these are considered to be "technical matters" that properly belong to the heteronomous sphere.[71] This presumably means that environmental matters such as energy budgets, resource use, pollution control, nature conservation, recycling, and workers' safety would fall within the province of the State and need not concern citizens, at least in their autonomous activities (although Gorz envisages that citizens would use durable, convivial tools in their free-time activity).

Yet Gorz's own brand of ecosocialism is politically problematic in naively defining the activities of the State—most notably, the provision of basic needs, the determination of socially necessary labor, and environmental protection—as mere technical administration that lies outside the realm of ethics or normative discourse. Here, the insights of the Frankfurt School seem to have been forgotten. Indeed, Gorz provides an active endorsement of what Habermas has described as the "scientization of politics." As we saw in chapter 5, this process has led to the gradual ascendancy of a technocratic elite and the withering of the public sphere. This approach stands in stark contrast to the project of the Frankfurt School and the community self-management approach of ecoanarchism, both of which seek full democratic participation in political, economic, technological, and ecological decision making.

Not surprisingly, Gorz's particular post-industrial utopia has attracted a number of criticisms from ecoanarchists and other ecosocialists. According to Murray Bookchin, Gorz's technocratic post-industrial utopia is riddled with paradoxes in attempting to combine central planning with neighborhood self-help initiatives and worker self-management. In particular, Bookchin argues that Gorz promises the impossible—central planning without bureau-cracy—but "tells us virtually nothing about the administrative structures around which his utopia will be organized."[72] In a similar vein, Richard Swift asks, what will prevent the heteronomous sphere or State from becom-ing "a center for the centralization of power? The tools for political abuse remain here."[73] Moreover, Gorz does not specify what role, if any, the mar-ket would play in his dual economy.

Other ecosocialists, however, are much clearer on the question of administrative structures and the role of the market. Generally speaking, most ecosocialists have embraced political pluralism, public accountability, and widespread public participation in economic planning. Moreover, they argue that democratic social planning would serve as the predominant resource allocation mechanism with markets playing a subsidiary role (e.g., in the small business sector). These general outlines of an ecosocialist econo-my, should not, however, be confused with "market socialism"—a predomi-nantly market system of exchange accompanied by social and/or State (as distinct from private) ownership of the means of production. Indeed, Boris Frankel has rejected "market socialism" (at least the kind advocated by Alec Nove in *The Economics of Feasible Socialism*) on the grounds that ecologi-cal and social justice objectives would be continually compromised by national and international market forces.[74]

EVALUATION: MORE DEMOCRACY OR MORE BUREAUCRACY?

The claimed superiority of the general ecosocialist economic program is that the State would no longer be fiscally parasitic on private capital accumula-tion to fund its social and ecological reforms. The upshot is that the contra-dictions of the market would be, for the most part, eliminated rather than simply "managed." The central question arises, however, as to whether the ecosocialist alternative can avoid the kinds of problems that have beset exist-ing command economies. These problems include bureaucratic corruption and bribery; underemployment of labor; gross economic inefficiencies; one-party dictatorships; intolerance toward political dissent; widespread political intimidation and oppression; economic stagnation, and ecological devasta-tion. In this new, post–Cold War era, ecosocialists must convince an increas-ingly skeptical public that it is possible to deliver economic planning that is at once democratic, ecologically responsible, coherent, and responsive to consumer demand.[75]

Now a *theoretical* case might still be made that a democratically planned economy is superior to a market economy in that it should be better able to (i) provide goods and services on the basis of need rather than purchasing power; (ii) avoid or minimize the "negative externalities" of a market econo-my; (iii) iron out excessive social and regional inequalities; (iv) ensure that the scale of the macro-economy respects the carrying capacity of ecosystems (unlike a market economy, a planned economy has no inbuilt imperative to grow); and (v) generally take a broader and more long-term view of the collective needs of present and future generations of both humans and nonhumans (i.e., unhampered by the need to appease the immediate interests of private capital).

If could be further argued that the ecological and social problems that have beset existing command economies can be attributed to a range of interrelated factors that are neither necessary nor desirable aspects of a democratically planned economy. These include rigid centralized control, single party bureaucratic rule, the absence of a free flow of information, the absence of an informed citizenry and popular participation, a commitment to industrialization and high growth rates, and a determination to "catch up" with the West in terms of technical development and military might as part of a perceived need to bolster "national security."[76] As O'Connor explains, "in all socialist countries the major means of production are nationalized although not yet socialized, i.e., there is no strong tradition of democratic control of the means of life."[77]

According to this argument, twentieth century communism must be seen as an aberration rather than as an example of the inherent tendencies or "logic" of a planned economy. If we remove all the repugnant features of these economies (as identified above) and ensure that production is properly "socialized" rather than simply "nationalized," then, so the argument runs, democratic self-management can emerge as a reasonably feasible option. As previously noted, a planned economy does not need to grow in the way that a market economy does. Moreover, we have seen that ecosocialists have explicitly rejected the path of indiscriminate economic growth and have embraced political pluralism, public accountability, freedom of information, and widespread public participation in economic planning.

Yet even if we confine our attention to this appealing theoretical case for a democratically planned economy, we can identify a number of *other* problems that might considerably undermine its apparent advantages. Just as the theoretical defence of the market is based on the extremely restrictive and unreal set of assumptions of perfect competition, the theoretical defence of a democratically planned economy is based on the unreal assumptions of, inter alia, full information and complete trust between principal and agent. As Jon Elster and Karl Ove Moene observe, "central planning would be a perfect system, superior to any market economy, if these two resources [full information and complete trust] were available in unlimited quantities."[78]

It is the lack of these two important "resources" that accounts for much of the interagency competition, corruption and bribery, displacement of responsibility, and political intimidation that have characterized the state agencies of existing centrally planned economies. Of course, these problems are not unique to bureaucracies in centrally planned societies. They are, however, magnified by virtue of the expanded role played by state agencies in a planned economy.

The theoretical defence of democratic economic planning also assumes that it is possible to coordinate successfully a range of different public agencies in accordance with a "common" economic plan. This, in turn, presupposes (i) that it is possible to arrive at a social consensus on a common economic plan; and (ii) that each agency charged with implementing aspects of the plan will interpret it in a uniform way. As John Dryzek points out, a teleological or goal directed social and economic system requires not just a consensus on values, but a *continuing* consensus on values if it is to produce, consistently and effectively, the desired common good. This arises from the fact that "the administered structure cannot waver in its commitment, for it is only that commitment which can keep the system on course."[79] This tells against the feasibility of a well functioning *and* democratic planned economy. It also sheds considerable light on the propensity of communist regimes to regularly intimidate—and, from time to time, eliminate—opposition to the planning dictates of the State.

Yet if participatory democracy, political pluralism, and the protection of civil liberties are to remain nonnegotiable elements of ecosocialism, then how can a popular consensus for a common economic plan be achieved, let alone maintained? (Classical liberal theorists, of course, have never had to wrestle with this dilemma, in that they have generally been more concerned with *procedural* fairness than with *end-state* fairness.) Moreover, even if we assume perfect information and perfect trust (two unlikely assumptions), how can economic information be processed and how can a *coherent* economic plan be successfully implemented and coordinated in the absence of a shared ecosocialist morality or, failing that, a controlling managerial elite?

The great strength of the price mechanism as a method of resource allocation, despite its many problems, is that it provides a relatively *decentralized* and *depoliticized* method of information processing and resource allocation that is responsive to consumer preference (at least in competitive markets). Its very "invisibility" (that is, the absence of a deliberative steering system or coordinating center) has helped to maintain its political legitimacy to the extent that it obviates the need for ongoing social debate and consensus as to the merits or demerits of resource allocation decisions. This recognition should not be taken as an endorsement of deregulation. Indeed, I have already outlined some of the more obvious ecological and social problems generated by market economies and accepted, at a very minimum, the need for State intervention to prevent or redress such problems. What is at issue here is the extent and character of intervention (whether by way of regulation or planning) and the implications this has for economic efficiency and democracy.

By comparison to predominantly market economies, predominantly planned economies are more visible, discretionary, and therefore more contestable than the impersonal, self-adjusting price signals of the market. In short, democratic collective planning necessarily attracts more criticism and debate as to the *desirability* of alternative courses of action. This is not necessarily a bad thing—indeed, it is precisely the kind of debate that ecosocialists wish to substitute for the impersonal and ethically blind signals of the market. Nonetheless, ecosocialists are unlikely to achieve the kind of *continuing* social consensus around a common plan that would be needed for a well functioning, planned economy.[80] Of course, these sorts of problems are not lost on ecosocialists. For example, the authors of *New Economic Directions for Australia* ask themselves and their readers the pertinent question: "How would the new planning processes and institutions form part of a coherent whole—more particularly how would they enshrine democratic principles and achieve the necessary balance between local, regional and national responsibilities?"[81] They also note that capital is likely "to move offshore the instant an Australian government begins to give the implementation of these proposals serious consideration."[82] Moreover, Ryle goes so far as to concede that the idea that the economy can be brought under democratic political control has "not been on the mainstream political agenda, and hence...[has] not been part of most people's sense of the possible, for many years now."[83]

The urgent task facing ecosocialists is to find ways of resolving the tension between their quest for participatory democracy, on the one hand, and coherent and efficient economic and ecological planning, on the other hand. That is, ecosocialists accept that the more the State intervenes in social and economic life, the more it needs to facilitate wide-ranging community consultation and consensus to maintain legitimacy. Yet the more the State replaces the market with a series of coherent economic and ecological plans, the more it also needs a central coordinating agency to ensure the successful implementation of such plans (and the more reluctant will be that central coordinating agency to encourage genuine democratic participation that might dispute the appropriateness, and block the smooth implementation, of such plans).

There are several ways in which this tension might be eased, if not resolved. For example, democratic planning would be considerably facilitated by a politically active, educated, and ecologically informed citizenry. Moreover, the move toward democratic economic planning would need to be implemented in gradual stages to allow for the discovery of, and adjustment to, unanticipated problems. This could be further faciliated by a multilayered political decision-making structure to enable a balanced democratic representation of local, regional/provincial, and national interests (this case concerning decision making structures is developed further in the following chapter).

Yet the more practical question still arises: do we have enough time, wisdom, and collective will to overcome the above tensions and move

toward a predominantly planned economy *to the degree* proposed by ecosocialists in view of widespread public skepticism toward economic planning
in the post–Cold War era and in view of the urgency of the ecological crisis?

AN ALTERNATIVE GREEN MARKET ECONOMY

A more feasible alternative might be to draw back from the idea of a predominantly planned economy in favor of the idea of a greater range of
macroeconomic controls on market activity that are designed to ensure that
market activity remain subservient to social and ecological considerations.
Here, the emphasis would be more on managing, containing, and disciplining
rather than largely replacing the market economy, although some economic
planning would still have a role to play. In suggesting that there might be at
least the outline of such an alternative, I will be drawing on the ideas developed by a small yet growing circle of economists associated with the New
Economics Foundation and TOES (The Other Economic Summit), who shall
be referred to for convenience simply as "Green economists" (although
strictly speaking, ecosocialist political economists represent one particular
school of Green economic thought).[84]

Although both ecosocialists and Green economists enlist the values of
participatory democracy, ecological responsibility, social justice, decentralization, and the dispersal of economic and political power, they differ over
how these values are to be interpreted and applied. Whereas ecosocialists
tend to emphasize the evils of the market economy and seek to democratize
and "ecologize" State and local economic planning institutions, Green
economists tend to emphasize the evils of central planning and seek to
democratize and "ecologize" the institutions of the market economy. However, both of these approaches look to the State (albeit in varying degrees and
for different purposes) to play a necessary coordinating, redistributive, and
planning role. Moreover, unlike the ecoanarchist current in Green political
thought, both of these approaches are decidedly post-liberal rather than anti-
liberal, insofar as they accept the Western liberal institutions of representative democracy, tolerance of political diversity, the rule of law and due process, and the protection of human rights (such as freedom of speech,
assembly, and organization).

Although Green economists are trenchant critics of corporate capitalism,
they are equally critical of the concentration of economic power in the hands
of the State. Accordingly, Green economists have tended to place greater
emphasis on the need to develop small business, local cooperatives, and local
economic self-reliance. Nonetheless, they also advocate increasing government economic management through "transformed-market conforming planning," that is, new institutional, fiscal, monetary, and pricing policies
designed to ensure that "the market has as intrinsic tendency to move in

directions that conform with the society's social and environmental goals."[85]
As Herman Daly explains, the general economic framework seeks

> to combine micro freedom and variability with macro stability and
> control. This means, in practice, relying on market allocation of an
> aggregate resource throughput whose total is not set by the market,
> but rather fixed collectively on the basis of ecological criteria of
> sustainability and ethical criteria of stewardship. This approach
> aims to avoid both the Scylla of centralized planning and the
> Charybdis of the tragedy of the commons.[86]

Rather than seeking a predominantly planned economy (with a small private
sector) Green economists seek to "bend and stretch" the "historically given
conditions currently prevailing."[87] In other words, they envisage a market
economy with a reasonably large private sector. However, they argue that all
market activity (whether carried out by the public or private sector) should
be more heavily circumscribed, scaled down (in terms of material-energy
throughput), and made more responsive to ecological considerations and
informed consumer preferences. This is consistent with the Schumacher-
inspired Green emphasis on human-scale institutions, decentralization, and
appropriate technology. As Daly and Cobb argue:

> If one favors independence, participation, decentralized decision
> making, and small- or human-scale enterprises, then one has to
> accept the category of profit as a legitimate and necessary source of
> income. There is plenty of room to complain about monopoly profits,
> but that is a complaint against monopoly, not against profits per
> se.... If one dislikes centralized bureaucratic decision making then
> one must accept the market and the profit motive, if not as a positive
> good then as the lesser of two evils.... We have no hesitation in opt-
> ing for the market as the basic institution of resource allocation.[88]

However, in defending the advantages of a market economy, Green
economists have no illusions about the present lack of consumer sovereignty
and the many contradictions, inequities, and negative externalities generated
by market rationality. Indeed, they have pioneered many aspects of the ecological critique of private and state capitalism, particularly the issues of the
scale of material and energy throughput and ecological carrying capacity.
Moreover, Green economists are highly critical of the New Right's case for a
"free" or thoroughly deregulated market. According to Paul Ekins, advocates
of the free market do not really want a free market; what they want is government protection of, and absence of interference with, *existing private
property rights*.[89] In contrast, the acceptance of the price mechanism by
Green economists does not carry with it an endorsement of the existing pattern of ownership, control, and wealth distribution nor the scale and reach of

market penetration in everyday life. In this respect, it is instructive to bear in mind the observations of Karl Polanyi on the different scale and role of the market in different historical epochs. According to Polanyi, "at no time prior to the second quarter of the nineteenth century were markets more than a subordinate feature of society."[90] In other words, it has only been *since* the industrial revolution that market relations have grown to dominate social life (rather than being mere accessories of social life). According to Polanyi, the most decisive changes occurred when land and labor became commodities (changes that were *instituted* by the liberal State). This served to transform property relations which, in turn, facilitated a transformation of social and ecological relations. Green economists are concerned to develop new community initiatives and new institutional frameworks that will see a gradual replacement of the traditional notion of private ownership and control of land with the notion of community trusteeship and stewardship (through, for example, the development of conservation and community land trusts).[91]

At the macroeconomic level, most Green economists look to the State to play a key role in reversing the privatization of benefits and socialization of costs that characterize the market economy. This includes breaking down monopolies and "excessive bigness"; providing those public goods and services that are not provided by the market; avoiding or redressing negative externalities; redressing regional and macroeconomic imbalances; redistributing wealth via a guaranteed income scheme; and ensuring an appropriate scale of macroeconomic activity relative to the ecosystem and biosphere. As Daly and Cobb have emphasized, the price mechanism merely ensures an optimal *allocation* of scarce resources (an efficiency issue—and even this is considerably undermined by the lack of perfect competition in the real world), not an optimal *distribution* of resources (a social/ethical issue) or an optimal *scale* of resource use (an ecological/ethical issue).[92] Moreover, ecocentric Greens argue that the question of appropriate scale and carrying capacity must be determined with reference to the needs of present and future generations of both the human *and* nonhuman world.

Most Green economists have tackled the problem of scale (i.e., the protection of ecological carrying capacity) by advocating (i) a range of new fiscal measures (such as resource depletion quotas and higher resource taxes and pollution charges) designed to control resource depletion and reduce material-energy throughput;[93] (ii) more comprehensive, and longer-range, environmental impact assessment and technology assessment;[94] and (iii) the replacement of indiscriminate GDP statistics with an alternative index of economic progress designed to provide a more meaningful yardstick by which to measure economic well-being. More generally, Green economists are seeking to shift the burden of taxation away from labor (an increasingly abundant "resource") and toward what are becoming increasingly scarce factors of production, namely, land, natural resources, and fossil fuel energy.

This will not only encourage greater efficiency in the use of these scarce factors but also encourage a greater use of human skills and thereby slow down the rate of substitution of labor by capital-intensive machinery.[95]

One convenient way of drawing together some of the major features of a Green market economy is to identify the kinds of changes that are sought in respect of the major links in the money cycle (i.e., savings, investment, and consumption). The major Green initiatives relating to savings and investment are the development of local credit and banking facilities, "ethical" investment funds, and self-managing local enterprises. Green economists also support the use of market-based incentives (such as pollution standards, charges or taxes, and in some cases marketable permits) that will ensure that prices reflect the true cost of production and consumption. Ideally, this means ensuring that there is no divergence between the private costs of production and consumption and the social and environmental costs of production and consumption.[96]

Some Green economists have also developed proposals for the reform of corporations to enable greater worker and community participation in investment decisions and more extensive worker and community ownership of capital assets. For example, Shann Turnbull has developed a range of new institutional reforms (including Ownership Transfer Corporations, Cooperative Land Banks, and Producer-Consumer Co-operatives) that will enable the development of "social capitalism."[97] Instead of relying on the public sector to redistribute wealth, social capitalism provides for the redistribution of the ownership of capital assets and the payment of a "social dividend" that would provide a guaranteed minimum income to all.[98]

The restoration of "consumer sovereignty" (through the cultivation of the well-informed, discerning Green consumer) provides a key plank in the strategy for a scaled-down, Green market economy.[99] In order to transform the countless individual acts of consumption into conscious "economic votes" that will influence investment decisions, Green economists argue for the development of independent consumer organizations that can keep consumers informed not only with regard to such matters as the price, quality, origin, and safety of products but also the social and environmental costs. Other proposals include stricter controls on labelling and advertising and, in some cases, the development of an independent ecological certification system.

Alongside the "formal economy," Green economists also stress the importance of nonmarket exchanges such as local barter, voluntary community services, and neighborhood reciprocity, all of which meet many needs that are unfulfilled by the State or the market. One of the purposes of the guaranteed minimum income scheme is to reduce peoples' dependence on paid work and increase the opportunity for people to engage in nonmarket exchanges and perform what James Robertson has referred to as "Ownwork" (i.e, self-organized work).[100]

To be sure, the Green economic program is not without its problems, the most challenging of which is the attempt to zealously pursue a policy of redistributive justice while simultaneously encouraging a contraction in the scale and rate of material throughput in a predominantly market economy. The success of such a program will largely turn on the extent to which a "Green State" will be able to facilitate a shift away from growth in material and energy throughput and toward qualitative or "post-material" growth in accordance with a new Green index of economic welfare. Moreover, although Green economists seek to retain and discipline rather than replace the price mechanism and private profit, they nonetheless confer on the State a considerably expanded range of economic powers, many of which are indistinguishable from those to be conferred on the ecosocialist State. This means that the democracy/efficiency tension will also be encountered, although less so than it would in an ecosocialist economy.

A case might also be made that Green economists have underestimated the cunning of market rationality in finding ways of circumventing their proposed new range of macro-economic controls. In this respect, Green economists would do well to direct more systematic attention to matters dear to the heart of ecosocialists, such as the reorganization of finance, credit, production, corporations, and work. Pursuing policies that are designed to achieve a more equitable distribution of the ownership and control of capital assets alongside a more democratic management of such assets would certainly relieve the redistributive burden (and size) of the State while also attracting the likely support of organized labor.

Finally, there are few material (as distinct from moral) incentives for exemplary ecological action—whether on the part of transnational corporations or nation States—in the competitive environment of global capitalism. Without concerted ecodiplomacy resulting in a comprehensive array of treaties providing for macro-ecological controls and standards at the *international* level, Green economists will remain hard pressed to convince an effective majority of voters within their own nation that they must become ecological saints while individuals and corporations in other countries continue to engage in ecologically irresponsible practices.

CHAPTER 7

Ecoanarchism: The Non-Marxist Visionaries

INTRODUCTION

The previous three chapters examined emancipatory theories that have either drawn upon, or otherwise maintained some general sympathy or continuity with, the extensive heritage of Marxism. This chapter examines a body of emancipatory theory that defines itself, by and large, as a distinct alternative to, rather than as an extension, reformulation, or revision of, this Marxist heritage.[1] This body of Green political theory is ecoanarchism.

This chapter identifies and evaluates two general currents of ecoanarchist thought—social ecology and ecocommunalism (the latter includes bioregionalism and what I call "ecomonasticism"). Although social ecology and ecocommunalism differ in many important respects they both share the following features. First, as the overarching anarchist categorization indicates, both seek to bypass and/or abolish the modern nation State and confer maximum political and economic autonomy on decentralized local communities. Second, both argue not only that anarchism is the political philosophy that is most compatible with an ecological perspective but also that anarchism is grounded in, or otherwise draws its inspiration from, ecology. Third, the theorists discussed in this chapter not only oppose all forms of social domination but also oppose, in varying degrees and for varying reasons, the domination of the nonhuman world. In particular, all ecoanarchists ground their political theory on an ecological perspective that seeks to transcend the resource conservation, human welfare ecology, and utilitarian preservation perspectives to which the ecophilosophical perspectives of eco-Marxism (including Habermasian Critical Theory) and ecosocialism are limited. Fourth, all of the ecoanarchist theorists discussed in this chapter are strong defenders of the grassroots and extra-parliamentary activities of the Green movement. Fifth, and finally, all of these theorists emphasize the importance of maintaining consistency between ends and means in Green political praxis.

Beyond this, however, ecoanarchists divide in terms of the various theoretical explanations they offer to account for the ecological crisis (e.g., social ecology attaches greater theoretical importance to social hierarchy than does ecocommunalism); the types of ecocommunities they advocate (e.g., social ecology is more libertarian, whereas the ecomonastic strand of ecocommunalism tends to be relatively more ascetic); and the *degree* to which they are critical of the Western anthropocentric heritage (e.g., the ecocommunal tradition is generally more ecocentric than social ecology). A further difference is that social ecology is largely the work of one particular theorist—Murray Bookchin—who has developed a distinctive organismic ecophilosophical perspective whereas ecocommunalism is a more general category that I employ to encompass a variety of other kinds of ecoanarchist approaches of a relatively more ecocentric persuasion.

THE SOCIAL ECOLOGY OF MURRAY BOOKCHIN

Murray Bookchin stands as one of the early pioneers of Green political theory. Over the past three decades, Bookchin's numerous publications on "social ecology" have sought to restore a sense of continuity between human society and the creative process of natural evolution as the basis for the reconstruction of an ecoanarchist politics.[2] Bookchin describes his thought as carrying forward the "Western organismic tradition" represented by thinkers such as Aristotle, Hegel, and, more recently, Hans Jonas—a tradition that is process oriented and concerned to elicit the "logic" of evolution.[3] According to Bookchin, the role of an ecological ethics is "to help us distinguish which of our actions serve the thrust of natural evolution and which of them impede it."[4] For Bookchin, evolution is developmental and dialectical, moving from the simple to the complex, from the abstract and homogenous to the particular and differentiated, ultimately toward greater individuation and freedom or selfhood. Social ecology—a communitarian anarchism rooted in an organismic philosophy of nature—is presented as the "natural" political philosophy for the Green movement because it has grasped this "true" grain of nature and can promise greater freedom or "selfhood" for both nonhuman nature *and* society. For Bookchin, an anarchist society, free of hierarchy, is "a precondition for the practice of ecological principles."[5]

Although ecosocialism and social ecology represent quite distinct schools of emancipatory thought, there is nonetheless a superficial resemblance between the two. For example, it is noteworthy that Bookchin has become a major voice of the "Green Left" in North America, sharing many of the European ecosocialists' criticisms of deep ecology (this is not surprising, given Bookchin's familiarity with Continental political theory and his former Marxist leanings). This shared critique arises from the fact that both social ecology and ecosocialism emphasize, as their names might suggest, the *social* origins of environmental degradation—an emphasis that has led both to criticize the deep ecology focus on anthropocentrism for deflecting attention away from inequities such as those based on class, gender, and race.[6] It is also interesting to note that Bookchin has claimed that the West German Green party "has supplanted the traditional socialisms with a social ecology movement."[7] As we saw in chapter 6, many ecosocialists also claim that the West German Green Party is essentially an *ecosocialist* party. The many factions within the West German Green party are such that both claims have some degree of truth.

Beyond these superficial similarities, however, social ecology departs fundamentally from ecosocialism in terms of both analysis and ecophilosophical perspective. Whereas ecosocialists have singled out capitalism as the main driving force behind environmental degradation, Bookchin has conducted a more wide-ranging critique that regards capitalism as but a subset of a more deep seated problem, namely, social hierarchy. Bookchin's social ecology, then, is not Marxist but rather *libertarian*. As Bookchin explains:

> To create a society in which every individual is seen as capable of participating directly in the formulation of social policy is to instantly invalidate social hierarchy and domination. To accept this single concept means that we are committed to dissolving State power, authority, and sovereignty into an inviolate form of personal empowerment.[8]

Bookchin is also critical of what he regards as the productivist and authoritarian legacy of socialist thought and its lack of a thoroughgoing ecological perspective.[9] Indeed, Bookchin's critique of André Gorz's ecosocialism is no less vehement than some of his recent criticisms of deep ecology. Despite Gorz's professed aim of going beyond Marxism, Bookchin argues that Gorz has merely fused Marxism and environmentalism, transcending neither, in a "politics that is environmentally oriented, not an ecological sensibility that is meant to yield a political orientation."[10]

Bookchin argues that social ecology, in contrast, *is* grounded in an ecological sensibility that rejects the instrumental posture toward nature that is characteristic of socialist thought. This follows from Bookchin's organismic philosophy, which recognizes "subjectivity" as present, however germinally, in all phenomena, not just humans. (By "subjectivity" Bookchin means any kind of purposive activity or striving, whether latent or advanced.[11]) Indeed, prior to Bookchin's much publicized attack on deep ecology at the National Greens Conference in Amherst, Massachusetts, in July 1987, many ecopolitical theorists regarded social ecology and deep ecology as complementary ecophilosophies. Both Naess's deep/shallow ecology distinction and Bookchin's social ecology/environmentalism distinction are critical of scientism and a purely instrumental orientation toward the nonhuman world; both seek to re-embed humans in the natural world; and, at a more practical level, both support bioregionalism, small-scale decentralized communities, cultural and biological diversity, and "appropriate" technology.

Despite these important commonalities, however, there are also important differences in the ecophilosophical orientations of deep ecology and social ecology that have given rise to different perspectives concerning humanity's proper role in the evolutionary drama. Moreover, since Bookchin's much publicized critique of deep ecology, these differences have become much more discussed, and hence much more marked, than the many commonalities, although there are more recent signs of a significant *rapprochement*.[12] Not only are important ecophilosophical issues at stake here; there is also a struggle to influence the political priorities of the growing Green movement. What, then, are the distinctive claims of social ecology?

Bookchin's Social Hierarchy Thesis

From as early as the 1960s Bookchin has maintained the thesis that the domination of nonhuman nature by humans arose from the domination of humans by humans. This argument finds its most developed expression in Bookchin's magnum opus *The Ecology of Freedom: The Emergence and Dissolution of Hierarchy*.[13] In this work, Bookchin has sought to develop this thesis by tracing the emergence of hierarchy and domination in human societies from the Paleolithic Age to modern times. According to Bookchin, the breakdown of "organic" or preliterate communities based on kinship ties into hierarchical and finally class societies, culminating in the State, was to gradually undermine the unity of human society with the nonhuman world. Bookchin argues that incipient domination arose originally in preliterate societies in the form of social hierarchies rooted in age, sex, and quasi-religious and quasi-political needs. These social hierarchies are presented as providing the conceptual apparatus of domination or what Bookchin calls the "epistemologies of rule"—the repressive sensibility of command and obedience that has enabled some humans to see others as objects of manipulation. According to Bookchin, "from this self-imagery, we have extended our way of visualizing reality into our image of 'external' nature."[14] This position has been reaffirmed in his more recent writings, where he argues that

> the domination of nature first arose within *society* as part of its institutionalization into gerontocracies that placed the young in varying degrees of servitude to the old and in patriarchies that placed women in varying degrees of servitude to men—not in any endeavor to "control" nature or natural forces. Various modes of social institutionalization, not modes of organizing human labor (so crucial to Marx), led to domination. Hence, domination can be removed only by resolving problematics that have their origins in hierarchy and status, not simply class and the technological control of nature.[15]

Bookchin's thesis that the domination of "external nature" by humans stems from the domination of some humans by other humans is a reversal of the general Marxist reading of history (welcomed by orthodox Marxists as a necessary stage toward communism and lamented by the first generation of Critical Theorists as repressive) that it is the increasing human mastery of nature that has given rise to class society and social domination. According to Bookchin, this Marxist reading of history saw the domination of nature as wedded to human survival in a hostile natural world. What was seen as humanity's ascent toward freedom from material want therefore demanded struggle, conquest, and our increasing disembeddedness from nature in order to overcome natural necessity. Social ecology, in contrast, does not see the conquest of nature as the necessary "price" of human freedom. Rather, it looks to nature as the *ground* of freedom and seeks to re-embed humans in the natural world.[16] This, argues Bookchin, demands the creation of a libertarian, stateless society that is guided by Bookchin's description of the ecosystem: "the image of unity in diversity, spontaneity, and complementary relationships, free of all hierarchy and domination."[17] This translates politically into a society that is free of "gerontocracies, patriarchies, class relationships, elites of all kinds, and finally the State, particularly in its most socially parasitic form of state capitalism."[18]

The radical thrust of social ecology is seen to arise from its ecological sensibility, which recognizes what is seen to be the nonhierarchical interdependence of living and nonliving things. According to Bookchin, such an ecological sensibility implicitly undermines the conventional notion that hierarchy is part of the "natural" order of things in both nature *and* society.[19] In the case of the nonhuman world, he argues that "the seemingly hierarchical traits of many animals are more like variations in the links of a chain than organized stratifications of the kind we find in human societies and institutions."[20]

Although Bookchin warns that we must avoid anthropomorphic judgements that project the distinctive features of socially created institutions onto the nonhuman world, he considers that the reverse is both legitimate and desirable. That is, we may "transpose the nonhierarchical character of natural ecosystems to society," provided we do not reduce social ecology to natural ecology.[21] This one-way mapping of ideas flows directly from Bookchin's organismic philosophy of nature, according to which humans are "nature rendered self-conscious," able to discern the thrust of evolution. In short, Bookchin argues that social hierarchy is undermined once we grasp that there is no hierarchy in nature.[22] He is concerned to criticize any social structure that inhibits self-determining activity, which is Bookchin's highest norm, since it coincides with what he takes to be the thrust of evolution. The gist of Bookchin's argument seems to be that by ridding society of social hierarchy (as distinct from differentiation or complementary relationships), we will remove the possibility of us developing hierarchical sensibilities vis-à-vis ourselves *and* the nonhuman world.

Does Bookchin's social hierarchy thesis offer a plausible explanation of, and solution to, social domination and the ecological crisis? Bookchin's thesis can be interpreted in a number of ways. An emphatic version of his thesis might be that social hierarchy *necessarily* inhibits the free unfolding of the human and nonhuman worlds; the corollary of this is that self-determining activity is possible only in a nonhierarchical society. Such an argument, however, is difficult to sustain. In human societies, social hierarchy can either enable or oppress, depending on the circumstances. Bookchin himself concedes that the existence of hierarchy or ranking in the form of social elites is not *necessarily* exploitative, at least in relation to so-called organic societies, for he states that

> the appearance of a ranking system that conferred privilege on one stratum over another, notably the young over the old, was in its own way a form of compensation that more often reflected the egalitarian features of organic society rather than the authoritarian features of later societies.[23]

Indeed, it is possible to envisage circumstances where social hierarchy can provide meaning, identity, and a context that *facilitates* personal self-realization. Take, for example, a monastery in which members may at any time renounce their vows and leave but who nonetheless voluntarily submit themselves to the authority of the abbot or Zen master. Here, the members are clearly self-determining individuals in the sense that they have freely chosen a disciplined lifestyle of obedience to a superior as a means of facilitating their own self-realization. Indeed, the monastic paradigm (discussed later in this chapter) has been widely referred to by other ecoanarchist theorists such as Bahro and Roszak as providing an exemplar of ecocommunal living by providing social solidarity, identity, and ecological harmony. Of course, this ecocommunal model of living does not necessarily require a guru, master, or governing committee at its apex. However, such a hierarchical arrangement, when accompanied by the qualities of personal example, integrity, and leadership (both spiritual and otherwise) in those who "rule," is quite acceptable to the ecocommunal tradition.

Of course, there are countless examples of human societies where social domination can be attributable to some kind of social hierarchy. Indeed, it would seem plausible to argue (and this may be all Bookchin intends to argue) that social hierarchy is merely a necessary (as distinct from sufficient) condition for social domination. In this case it is not necessary to show that social hierarchy will *always* give rise to social domination in every case—only that it always be present in those cases where social domination does exist and that social domination will never be found in those societies in which social hierarchy is absent. Even here, however, one might want to consider the positive benefits of social hierarchy (in, for example, tribal societies) and weigh these benefits against the potential downside rather than opposing social hierarchy in every instance.

While the argument that social hierarchy gives rise to *social* domination has general plausibility, the second aspect of Bookchin's social hierarchy thesis is much more contentious, namely, that the root of the *ecological crisis* is social hierarchy and, accordingly, the solution to that crisis is the removal of all forms of social hierarchy. Again, if by this argument Bookchin means that there is a necessary connection between social hierarchy and the domination of the nonhuman world, then his thesis can be simply refuted: first, by the many historical examples of hierarchical societies that have lived in relative harmony with the nonhuman world (e.g., Benedictine communalism, feudalism, many preliterate societies) and, second, by the theoretical possibility of an egalitarian, nonhierarchical society that nonetheless continues to dominate the nonhuman world. As to the latter, Marx's communist society of the future would have been a clear example of this had it eventuated. As Fox puts it, "Bookchin's presentation of social ecology thus conveys no real appreciation of the fact that the relationships between the internal organization of human societies and their treatment of the nonhuman world can be as many and varied as the outcome of any other evolutionary process."[24]

Now it is possible that this represents an oversimplified and overstated interpretation of the ecological aspect of Bookchin's case—indeed, it is often very difficult to pin down Bookchin's arguments with any degree of precision.[25] In view of this, it will be useful to examine a less emphatic and more qualified (and more plausible) version of his social hierarchy thesis as it relates to the nonhuman world. That is, it is possible to interpret Bookchin as saying only that history has shown that hierarchical societies create the psychological conditions for the domination of the nonhuman world but that the actual domination of the nonhuman world is dependent upon a society possessing the requisite technology. In this case, social hierarchy is merely a necessary, as distinct from a sufficient, psychological condition for the domination of the nonhuman world, just as it is merely a necessary, as distinct from a sufficient, psychological condition for social domination.

Yet even when Bookchin's thesis is expressed in these more qualified terms, it arguably has no *more* plausibility than the early Frankfurt School thesis that he reverses. As we saw in chapter 5, the early Frankfurt School theorists had argued that the domination of external nature through the development of more sophisticated and large-scale technologies (a development that resulted from the ascendancy of instrumental rationality over other kinds of rationality) has given rise to social hierarchy and the domination of people. Note that the crucial link in both theses is the presence of a certain mental framework or way of seeing ("epistemologies of rule" in the case of Bookchin, instrumental rationality in the case of the Frankfurt School); note also that in both cases the actual domination of the nonhuman world is dependent on this mental framework finding practical expression in repressive technologies. When broken down in this way, the qualified version of Bookchin's thesis and the early Frankfurt School thesis may appear more as *complementary* (rather than oppositional) theses that examine, as it were, different sides of the same coin.

Yet if it is accepted that technology is our principal mediator with the rest of nature, shaping how we interact with ourselves and the nonhuman world, then a case might be made that the early Frankfurt School's thesis has more plausibility than the qualified version of Bookchin's thesis. For example, the historical exploration of the oriental despotism of "hydraulic societies" by the Frankfurt School theorist Karl Wittfogel shows how large scale interventions in the "natural" flow of water through the mighty irrigation projects of ancient China, India, Mesopotamia, and Egypt necessitated large scale, totalitarian, and centralized bureaucracies that impoverished the lives of peasant populations.[26] More recently, the American environmental historian Donald Worster has employed the early Frankfurt School thesis to explain the highly centralized bureaucratic political order that accompanied the intensive irrigation of the Colorado River—a development that resulted in the farmers of the Colorado River basin becoming virtual wards of the massive United States Bureau of Reclamation.[27] A similar analysis might be applied to the "hydro-industrialization" policy of a succession of Tasmanian state governments from the 1920s to the mid-1980s, which resulted in the steady augmentation of the power of the State's Hydro Electricity Corporation to the point where it was, for a time, no longer answerable to Parliament for the conduct of its affairs.[28] In these and many other examples of large-scale technological interventions in nature—such as the modern day massive hydroelectric projects in India, Egypt, and Brazil—"'administered life' is coming to be the common experience for rivers, deserts, and people alike."[29]

Now Bookchin would probably argue that these examples support *his* thesis by pointing to a pre-existing social hierarchy in these societies that was able to envisage and implement these gigantic technological feats in the first place. Moreover, he would argue that if we are to overcome these domineering institutions, we must first change our social organization rather than our technologies because "a liberatory technology presupposes liberatory institutions."[30]

Yet we have already seen that it is at least theoretically possible to have a society that is free of social hierarchy but that nonetheless dominates the nonhuman world through large scale technologies in order to minimize "necessary labor" (such as Marx's communist society). In this case, human emancipation *requires* the domination of "external nature." Of course, Bookchin rejects this orthodox Marxist route to human emancipation and emphasizes the importance of democratically manageable technologies and "soft" energy paths—reforms that Bookchin argues will naturally flow from the kind of social reorganization he envisages. But surely such ecologically benign technologies would arise not simply from the absence of social hierarchy per se but rather from the *absence* of social hierarchy *plus* the *presence* of a certain kind of ecological sensibility. Indeed, as we saw in chapter 4, it is precisely the *absence* of a thoroughgoing ecological sensibility in the theoretical categories of orthodox Marxism that accounts for orthodox Marxism's instrumentalist and technocentric posture toward the nonhuman world.

These examples are not intended to dismiss Bookchin's case altogether but rather to point to alternative ways of addressing the relationship between social domination and the domination of the nonhuman world. Indeed, I would suggest that it is probably impossible to declare that either Bookchin or the early Frankfurt School is "right" or "wrong" in view of the chicken-and-egg nature of the problem we are examining. At a minimum, these alternative approaches serve to highlight the mutually reinforcing nature of the domination of the human and nonhuman worlds by some humans, at least in relation to large scale, technological interventions in nature. It also highlights the complexity and mutually reinforcing relationship between a society's social institutions and technologies and its orientation toward the nonhuman world. What I would like to explore further, however, are the different *political priorities* that flow from these opposing theses, given that it is possible to address the general problem of domination from different angles.

Both approaches arrive at broadly similar political conclusions, that is, the need to move toward a society operating on the basis of smaller scaled, decentralized, and ecologically benign technology and energy sources, and greater local democracy and social cooperation. Nevertheless, the political route by which such a society would be arrived at would be quite different in each case. Bookchin's thesis gives strategic priority to dismantling all hierarchical institutions within society and establishing a direct democracy at the community level; the resolution of ecological problems will come later by virtue of the general nonhierarchical sensibility that an ecoanarchist society would engender, a sensibility that Bookchin argues would be extended to the nonhuman world. In contrast, the early Frankfurt School thesis sought an end to scientism and the dominance of instrumental reason. Applied in the context of the modern ecological crisis, this thesis would maintain that opportunities for greater individual autonomy in society will flow from a reorganization of society's scientific and administrative institutions and technologies along more democratic and less "imperialistic" lines. This would lead to the breakdown of bureaucracies, the decentralization of energy resources, and the development of a more socially and ecologically responsible science. This, in turn, would enable the liberation of "repressed human sensibilities," to adopt the language of the early Frankfurt School.

Of course, Bookchin has also argued on many occasions for the cultivation of an ecological technics as well. However, in terms of his *theoretical* analysis, this would not be given the same *priority* as removing social hierarchy because the widespread use of ecological technics is seen to *depend* upon the removal of social hierarchy. From an ecocentric perspective, the urgency of the ecological crisis and the alarming rate of species extinction demand that we give top priority to employing technologies that maintain ecosystem viability for both the human and nonhuman worlds. Now ecocentric theorists would agree with Bookchin that the revival of local democracy and the breakdown of excessive concentrations of economic and political power (e.g., in bureaucracies and corporations) are essential to social emancipation (just as Bookchin accepts the need for an ecological technics). Yet, as Lorna Salzman argues, "a radical, ecologically inspired-politics that aims at ecological sanity and reconstruction *necessarily subsumes all the issues of socioeconomic injustice and oppression* with which social ecologists are concerned."[32] For example, the straightforward decision by a democratically elected parliament to move toward a decentralized and predominantly solar- and wind-powered local economy would have the effect of breaking down the economic and political power of large scale, centralized bureaucracies and corporations. It would also change the scale and nature of institutions in such a way as to make them more amenable to participatory democracy.

Finally, as we shall see, social emancipation and "ecological emancipation" are not incompatible with the continued existence of the State. Quite the contrary, both can be facilitated by the State. Indeed, the urgency of the ecological crisis is such that we cannot afford *not* to "march through" and reform the existing institutions of liberal parliamentary democracy (where they are available and despite their many limitations) and employ the resources (legal, financial, and diplomatic) of the State to promote national and international action, curb ecological degradation, and foster the redistribution of resources between the rich and poor nations of the world.

Bookchin's Evolutionary Stewardship Thesis

Bookchin's anarchist ideal of freedom is one that sees any kind of other- or external-directedness, as distinct from self- or internal-directedness, as thwarting impulses that are deemed to be natural and good. Like many anarchists, Bookchin enlists evolutionary theory to support the notion of the inherent sociality of humanity and to claim that any form of higher or external authority is *against nature*. In this respect, George Woodcock's general observations on anarchism are pertinent to social ecology (although Bookchin speaks of evolutionary "directionality" rather than "natural laws"):

> The whole world-view within which anarchism is embraced depends on an acceptance of natural laws manifested through evolution, and this means that the anarchist sees himself as the representative of the true evolution of human society, and regards authoritarian political organizations as a perversion of that evolution.[33]

Yet it is problematic to invoke a presumed telos in nature as a justification for social ecology or indeed any political theory, since even if we assume that such a telos accorded with the picture of nature provided by modern science (an assumption that is open to serious challenge) this does not in itself tell us why we *ought* to follow it. As J. Hughes observes, "the reason for us to oppose hierarchy has to do with an existential human ethical decision, not with its 'unnaturalness.'"[34]

Bookchin's teleological approach departs considerably from the more agnostic approach adopted by ecocentric theorists regarding the direction of evolution. As we have seen, Bookchin's organismic philosophy emphasizes the active, creative role played by humans in the evolutionary drama (indeed, he often refers to humans as "nature rendered self-conscious"). Moreover, he emphasizes humanity's creative role in natural evolution by distinguishing between what he calls "first nature" (the human world) and "second nature" (the nonhuman world), which is intended to capture the *graded* development of the human world out of the nonhuman world. Second nature, according to Bookchin, enables first nature to act upon itself *rationally*:

> We cannot hope to find humanity's "place in nature" without knowing how it *emerged* from nature with all its problems and possibilities. Our result yields a creative paradox: second nature in an ecological society would be the actualization of first nature's potentiality to achieve mind and truth.[35]

This creative role assigned to humans in fostering nature's evolution is the essential basis upon which Bookchin rejects asceticism, stoicism, biocentrism/ecocentrism, or any worldview that he interprets as involving the "quietistic surrender" or resignation by humans to the natural order. Bookchin interprets such approaches (quite wrongly, in the case of ecocentrism) as idolizing and reifying nature and setting it apart from a "fallen humanity"—an approach that Bookchin claims is an insult to humanity by denying us our creative role in evolution.[36] There must be an infusion of human values into nature, he argues, because humans are the fulfillment of a major tendency in *natural* evolution. Indeed, Bookchin claims that our uniqueness cannot be emphasized too strongly "for it is in this very *human* rationality that nature ultimately actualizes its own evolution of subjectivity over long aeons of neural and sensory development."[37] The clear message of Bookchin's ethics, then, is that humanity, as a self-conscious "moment" in nature's dialectic, has a responsibility to rationally direct the evolutionary process, which in Bookchin's terms means fostering a more diverse, complex, and fecund biosphere. Indeed, he suggests that we may "create more fecund gardens than Eden itself."[38]

Bookchin's view of humans as evolutionary stewards is at odds with an ecocentric orientation toward the world in two important respects. First, an ecocentric orientation is more concerned with "letting things be," that is, allowing all beings (human and nonhuman) to unfold in their own way (the means of achieving this goal, of course, require *active* political engagement in defence of the Earth and oppressed peoples). From a long term ecological and evolutionary perspective, adaptation, change, innovation, destruction, and extinction are recognized as features of natural systems, but rather than being fostered or accelerated, they are allowed to unfold in accordance with natural successional and evolutionary time. This is because ecocentric theorists do not see later, more developed, or more complex life-forms as necessarily "higher" or "better" than earlier, less developed, or more simple ones. Second, an ecocentric perspective adopts a more *humble* position than social ecology insofar as it does not claim to know what the thrust or direction of evolution is. Rather, an ecocentric perspective remains open minded toward what is seen as an essentially open ended process. As Lorna Salzman points out:

> Extinction of species has been a fact—a second species of *homo* co-existed with *h. sapiens* until relatively recently. The fact that we are (or believe we are) the only self-aware species on Earth (which we cannot prove) does not mean that this was evolution's impulse or our "striving." We need not have survived at all; there was and is no "necessity" that we do so.[39]

From an ecocentric perspective, it is both arrogant and self-serving to make, as Bookchin does, the unverifiable claim that first nature is striving to achieve something (namely, greater subjectivity, awareness, or "selfhood") that "just happens" to have reached its most developed form in *us*—second nature. Bookchin, in contrast, insists that "'Gaia' and subjectivity are more than the effects of life; they are [expressions of] its integral attributes."[40] His philosophy of nature is predicated on the intuition that there *must* be a telos (in the sense of a general directionality as distinct from a fixed end) in nature by virtue of the wondrous patterns it reveals:

> from the ever-greater complexity and variety that raises sub-atomic particles through the course of evolution to those conscious, self-reflective life forms called human beings, we cannot help but speculate about the existence of a broadly conceived *telos* and latent subjectivity in substance itself that eventually yields intellectuality.[41]

It is highly contentious, however, to ascribe a purpose to a particular development on the basis of its outward results. Bookchin's speculations go well beyond the more parsimonious explanation for the evolution of human consciousness provided by modern evolutionary theory. Here, the basic Darwinian idea of natural selection (i.e., random mutation and "selective" environmental pressures) remains, despite many important additions and revisions, the cornerstone of modern evolutionary theory (although the emphasis has shifted from the evolution of an organism *against* an environment as background to the coevolution of an organism *with* its environment).[42] According to this picture,

> evolution is basically open and indeterminate. There is no goal in it, or purpose, and yet there is a recognizable pattern of development. The details of this pattern are unpredictable because of the autonomy living systems possess in their evolution as in other aspects of their organization.[43]

Of course, the current scientific understanding of evolution is surrounded by much controversy and is hardly complete. But in the face of this uncertainty it is noteworthy how often we select, *from a range of possible conclusions,* those conclusions that are most comforting to humans. For his part, Bookchin scorns any attempt to explain the development of natural phenomena as an accident: "To invoke mere fortuity as the *deus ex machina* of a sweeping, superbly organized development that lends itself to concise mathematical explanations is to use the accidental as a tomb for the explanatory."[44]

But, as Fox and other nonanthropocentric environmental philosophers point out, the challenging thing about science vis-à-vis our understanding of our place in the scheme of things is that it often delivers us news about the universe that we might not wish to hear (indeed, George Sessions has argued that modern science has "been the single most decisive nonanthropocentric intellectual force in the Western world").[45] In developing an ecocentric political theory, then, we must remain open to this "news" while at the same time recognizing our own fallibility and the contingency of scientific explanation. In particular, ecocentric theorists take seriously the possibility that nature is not only more complex than we presently know but also quite possibly more complex than we *can* know. The wisest course of action from an ecocentric perspective is not to presume that we know the thrust of evolution but instead to remain open minded and, wherever practical, simply "tread lightly" in the course of sustaining ourselves and our human society.

The different ecophilosophical orientations of social ecology and ecocentrism give rise to different emphases when it comes to the "litmus issues" of wilderness preservation and human population growth (although these different emphases have become less marked in the wake of recent debates between deep ecologists and social ecologists). For example, although Bookchin has often emphasized the value of spontaneity, it is noteworthy that, *by comparison to ecocentric theorists* (and prior to recent debates and reconciliations), Bookchin has written very little on the subject of wilderness preservation as compared to, say, issues concerning social organization and the urban and agricultural or "domestic" environment. Indeed, Bookchin has stated that he does not believe that human stewardship of the Earth "has to consist of such accommodating measures as James Lovelock's establishment of ecological wilderness zones." Rather, it requires "a radical integration of second nature with first nature along far-reaching ecological lines."[46] This is in keeping with Bookchin's interpretation of humans as ecological stewards, playing an active and creative role in the evolution of the planet. In contrast, it has been the intensive philosophical investigation of wilderness issues and the question of the moral standing of nonhuman life-forms and entities by many ecophilosophers that has contributed to the formulation and development of the modern ecocentric perspective. Although ecocentric theorists do not deny the creative evolutionary role played by humans, and although they would like to see more ecologically integrated human communities, they have *consistently* maintained that the preservation of large tracts of wilderness is presently one of the most reliable and appropriate means of protecting a threatened array of nonhuman life-forms (and a diminishing number of indigenous human cultures) from the seemingly insatiable resource demands of human economic development and an expanding human population. Indeed, ecocentric theorists are generally opposed to the idea of any further destruction of the remaining areas of wilderness throughout the world—a position that Bookchin has (*contra* earlier statements) recently endorsed in his public reconciliation with Dave Foreman of the Earth First! movement.[47] (Of course, ecocentric theorists also argue that wilderness protection needs to be accompanied by far-reaching social and economic changes designed to reduce overall environmental impact and enhance social equity.)

Consistent with their strong stand on the need for wilderness preservation, ecocentric theorists advocate a population policy that seeks a long term, gradual *reduction* in human numbers. This position flows directly from the ecocentric concern for biological and cultural diversity; from the ecocentric concern to provide space for all beings to unfold in their own ways. To the extent that Bookchin's approach to the population question can be discerned from his criticisms of deep ecology, his emphasis has been on the need to overcome social hierarchy, decentralize society, redistribute resources, and cultivate an "ecological technics" rather than on the need to address the problem of absolute numbers per se by means of birth control programs.[48]

On the population question, then, Bookchin has much more in common with ecosocialism than ecocentrism insofar as he points to social relations rather than absolute human numbers per se as the "real causes" of famine and environmental degradation. Indeed, Bookchin has gone much further than ecosocialism in criticizing ecocentrism as racist and misanthropic for ignoring inter-human inequities. While it is not my concern to present a detailed overview of Bookchin's accusations in this controversy, it is important to lay to rest one serious misconception of the ecocentric position that arises from Bookchin's critique. As should be clear from the discussion of the population question in the previous chapter, ecocentric theorists do not *only* advocate a long term, gradual reduction in absolute human numbers in response to the population issue, as some of their critics suggest. They also advocate a more equitable inter-human distribution of resources, a lower overall level of resource consumption per capita, and the introduction of ecologically benign technologies. Moreover, the charge by social ecologists and ecosocialists that the ecocentric analysis is Malthusian (i.e., famine is inevitable and/or necessary to preserve carrying capacity) is completely misleading since it conflates the debate concerning the relationship between human population, hunger, and food distribution (which is the basic Malthus/socialist controversy) with the debate concerning the relationship between human population growth as a whole and general environmental degradation. As to the former debate, an ecocentric perspective would necessarily reject the Malthusian response and support the case for a more equitable pattern of control and distribution of resources among the world's existing human population alongside the development of more "appropriate" food sources and production methods. This is because an ecocentric perspective is concerned with human *and* nonhuman emancipation. As to the latter debate, however, the ecocentric position maintains that even if we assume that these social reforms would overcome the problem of hunger for the world's *existing* human population (an assumption some scientists would now dispute), it would not alone solve the more general ecology crisis for the existing nonhuman community or for future generations of humans and nonhumans. As we saw in the previous chapter, environmental impact is a function not only of technology and affluence (i.e., level of consumption) but also *absolute* human numbers. Accordingly, pollution, habitat destruction, and species extinction would continue apace as more intensive agriculture and industry expanded to meets the needs of an expanding population. This is why we cannot afford simply to await the "demographic transition."

According to projections by the United Nations Population Division, the world's human population is expected to increase to 8.5 billion over the next 35 years (to 2025), and 95 percent of the projected increase of 3.2 billion is expected to occur in the less developed countries.[49] This raises pressing international equity issues that threaten global environmental integrity and security. From an ecocentric perspective, it demands that the developed countries redistribute resources to the developing countries (to assist them in undergoing an ecologically benign demographic transition) while scaling down their own consumption of resources to a level that is compatible with global justice. But this alone is not enough. The ecocentric argument is that both human and nonhuman communities need space and a healthy and varied diet and environment in which to flourish; if this is accepted, then it is essential that we also address the problem of absolute human numbers through democratic family planning *in addition* to technological and inter-human distributional questions. This is a *non*anthropocentric population policy and it should not be confused (as ecosocialists and social ecologists are prone to do) with a misanthropic or racist one. As Arne Naess and George Sessions state in their deep ecology platform: "The flourishing of human life and cultures is compatible with a substantial decrease of the human population. The flourishing of non-human life requires such a decrease."[50] Indeed, Arne Naess has suggested that the "destruction of cultural diversity is partly the result of too many humans on earth" and that cultural diversity might therefore be enhanced if we move toward a gradual, long-term reduction in the human population.[51]

We may conclude from the foregoing exploration of Bookchin's ideas that social ecology is less anthropocentric and more ecologically inspired than all of the emancipatory theories examined so far in part 2, but that it is not an ecocentric emancipatory theory. Rather, social ecology may be seen as offering a radical ecological humanism that stands somewhere between ecosocialism and ecocommunalism on the anthropocentric/ecocentric spectrum. Does this mean that ecocommunalism represents the appropriate terminus of our political inquiry?

ECOCOMMUNALISM

Ecocommunalism is a generic term that encompasses a diverse range of utopian, visionary, and essentially anarchist Green theories that seek the development of human-scale, cooperative communities that enable the rounded and mutualistic development of humans while at the same time respecting the integrity of the nonhuman world. Progress, according to eco-communal theorists, is generally measured by the degree to which we are able to adapt human communities to ecosystems (rather than the other way around) and by the degree to which the full range of human needs are fulfilled. Like Bookchin, these theorists are generally critical of purely instrumental valuations of the nonhuman world and instead appeal to nature as a source of both inspiration and guidance.

The idea of building stable communities in accordance with the "principles of nature" is not new. According to Robert Nisbet, since the fall of the Roman Empire, the idea of an "ecological community" has been an enduring theme in Western social philosophy. It began with the monastic orders founded by Saint Benedict of Nursia following the collapse of the Roman Empire and has continued through the Utopian tradition in political thought, from Sir Thomas More's *Utopia* in 1516 down to the communal longings of the nineteenth and twentieth century anarchist philosophers (most notably William Godwin and the Russian Prince Kropotkin).[52] Writing in the early 1970s—well before the flowering of the Green movement and Green parties—Robert Nisbet defined the essence of the ecological community as

peaceful, not concerned with capture and forced adaptation, noncoercive, and seeking fulfillment through example or vision rather than through revolutionary force and centralization of power. The uncovering of those autonomous and free interdependencies among human beings which are believed to be natural to man and his morality: this—not the violent capture of government, army, and police—is the most fundamental aim of the tradition of community in Western social thought I call ecological.[53]

Nisbet defends his use of the adjective "ecological" to describe this tradition of thought on the grounds that (i) the Greek word *oikos* refers to the house-hold, and by implication, "to the natural and harmonious interdependencies of the household economy," (ii) in post-Darwinian usage, *ecology* refers to the natural interdependencies among organisms and between organisms and their environment, and (iii) in recent times ecology has taken on a *moral* overtone concerned with protecting "natural" interdependencies as distinct from those which are artificial or contrived.[54] Nisbet's characterization is one that merges the ideas of ecology as a science and ecology as a philosophy of life; it thus provides a statement of what the natural order is like (if left to unfold according to its own "laws") as well as a statement of how it ought to be.

Nisbet identifies five characteristics of the ecological community. The first of these is the idea of nature serving as a regulative ideal, with "nature" being employed in the Greek sense of what is "the normal, inherent constitution or manner of growth of an entity in time," unimpeded by alien impact, accident, or human evil.[55] The second is the notion of the "web of life"—a profound sense of our harmonious "relationship with other beings in the kingdom of life and of the necessity of maintaining this relationship, indeed of heightening it, through close contact with the land and all that grows on the land."[56] Humans were seen as part of a larger natural order that was divine, a conviction that was expressed in the Benedictine respect for the seasons and the soil and, more explicitly, in the sermons to birds, animals, the sun, and the moon given by Saint Francis of Assisi, the founder of the Franciscans. (It is noteworthy that the historian Lynn White has proposed St Francis as "a patron saint for ecologists").[57] Associated with the idea of a harmonious balance in nature is the insistence on developing a harmonious balance between thought and labor, factory and field, mind and body, and culture and nature—themes that were particularly dear to the hearts of nineteenth century ecocommunal and utopian theorists such as Peter Kropotkin and William Morris. A third feature of this tradition identified by Nisbet is a hostility to greed and competition and the fostering of community-based, cooperative modes of living and working. Cooperation or "mutual aid" (Kropotkin's oft-quoted phrase) was accepted as an essential part of the web of life and, according to Nisbet, the "highest ideal" of the ecological community. Fourth, such communities were also to be free from arbitrary authority, coercion, or repressive law. Nisbet argues, however, that they were not, except in their naive form, without order or discipline. This principle of "autonomous association" was well illustrated in the Benedictine Rule according to which membership of the monastery was conditional upon the member's desire to remain part of the community. For so long as a person remained a member of the community, however, authority would be asserted (by the abbott) in respect of errant members through the sparing applications of certain sanctions. Unlike Bookchin's social ecology, then, this tradition of the ecological community has countenanced certain localized forms of social hierarchy. Finally, Nisbet has found in this tradition of ecological communities "a clear and unwavering emphasis upon simplicity." Hyperorganization and complexity were condemned as working against the ideal of harmonious balance between mind, body, and spirit: "Nature, it is said, is simple for those who understand; society should be also."[58]

These five elements identified by Nisbet—nature as a regulative ideal, a profound respect for the web of life, cooperation, autonomous association, and simplicity—in may ways capture the essence of what I have identified as the ecocommunal strand of emancipatory thought that has emerged in the late 1970s and early 1980s. For example, all of these themes can (with varying degrees of emphasis and articulation) be found in the bioregionalism of Kirkpatrick Sale, the "small is beautiful" theme and "Buddhist economics" of Fritz Schumacher, the "Liberated Zones" of Rudolf Bahro's ecofundamentalism, and in Theodore Roszak's theme of "person/planet," to name four significant contributors to this body of Green thought.[59]

The most noteworthy feature of the modern incarnation of this tradition of "ecological community," however, is its general ecocentric orientation. Whereas liberalism and Marxism have generally regarded the domination of nature as the necessary price of human freedom, ecocommunal theorists argue that the cultivation of an attitude of respect for nature is a necessary aspect of human psychological maturity and self-realization. Ecocommunal theorists see humans as forming part of a larger (and to some, divine) order or process. A further notable feature of ecocommunalism is a concern to develop local socioeconomic arrangements that are geared toward the satisfaction and integration of the full range of human needs (that is, psychological, spiritual, intellectual, creative, and social needs, as well as material needs). Unlike, many Marxist and neo-Marxist theorists, ecocommunal theorists do not draw on the problematic distinction between freedom and necessity. True freedom or self-realization is not something that can be experienced only beyond "bread labor." Rather, it is a function of the extent to which an individual's entire range of needs are integrated and satisfied. What some liberals call "negative freedom" or "freedom from" (meaning the absence of constraints) is, for ecocommunal theorists, merely a condition for the realization of "positive freedom" or "freedom for," which is the rounded, mutual development or self-realization of the individual and ecocommunity.

Beyond these general points, however, there is a considerable variety of approaches in the ecocommunal tradition, a fact that is to be expected in view of its emphasis on diversity and local autonomy.

The utopian and anarchist dimensions of ecocommunalism have made it vulnerable to a range of criticisms: that it is naive, voluntarist, simplistic, and blind toward certain recalcitrant aspects of human nature. These are serious obstacles to the widespread acceptance of ecocommunalism as an appropriate political framework for social and ecological renewal in the modern world. Indeed, a case will be made that ecocommunalism needs to be supplemented by political engagement with state institutions if it is to not to remain an ephemeral and/or marginal political phenomenon. Moreover, the idea of nature as presented by some ecocommunal theorists is often anachronistic and/or idealized and in need of reformulation. This does not, however, invalidate the general orientation of this tradition and the importance of leading by example.

Before exploring the general limitations associated with the utopian and anarchist character of this tradition, it will be helpful to introduce some modern examples of the the ecocommunal current in Green political theory.

Monasticism Revisited

A common theme among ecocommunal theorists is the idea of disengagement or withdrawal from corrupt social and political institutions and the establishment of exemplary institutions and/or the pursuit of exemplary personal action. In view of the sweeping ecocommunal critique of most aspects of modern industrial society, such a strategy is, in many respects, the only reliable and authentic strategy that will maintain consistency between ends and means. In arguing for the establishment of ecological communities as the solution to the multifaceted crises facing modern society, a number of ecocommunal theorists have employed the analogy of the emergence of the medieval communalism of Saint Benedict of Nursia out of the ruins of the Roman Empire. This reference to the monastic paradigm as the nucleus for an ecological community is not just a recent phenomenon but can also be found in some strands of utopian socialism. William Morris, for example,

> wrote of setting up a brotherhood, a new monastic order for a "Crusade and Holy Warfare, against the age, 'the heartless coldness of the times.'" He read Thomas Carlyle's book, *Past and Present* (1843)—a glowing account of a twelfth-century monastery—with eagerness.[60]

Indeed Morris's own medievalist and romantically inspired social and political thought possessed all the characteristics of the ecological community. He rejected what he called "utilitarian sham Socialism" (which he felt was concerned merely with the organizational means for improving the material conditions of the working class) in favor of a community-based, "ethical social-ism" that coexisted in harmony with nature, cultivated the "whole person," and restored the dignity and creativity of labor as craft. (It is noteworthy that the romantic and utopian character of Morris's socialism—particularly his utopian novel *News From Nowhere*—has generated a considerable debate as to whether he properly qualifies as a Marxist.[61])

In contemporary ecocommunal literature, references to the monastic paradigm abound. For example, the ideal of "withdrawal and renewal" is central to Rudolf Bahro's ecofundamentalism, and he has often claimed that the establishment and growth of small-scale cooperatives or "Liberated Zones" (his ecocommunal solution to the ecological crisis) would lead society toward a better future in the same way that the communes founded by Saint Benedict were intended as a return to community and order after the chaos and social decay that had set in following the collapse of Rome.[62] Liberated Zones would ensure consistency between ends and means by providing both a supportive refuge from the destructiveness and alienation of industrialism and the nucleus of a new "biophile [i.e., life-loving] culture." Bahro is at pains to point out that the challenge of ecological degradation is primarily a cultural and spiritual one and only secondarily an economic one. Accordingly, we must direct our attention to cultural and spiritual renewal rather than structural or economic reform. Liberated Zones thus provide, in Bahro's view, a total solution to the multifaceted crises of modern times.

Similarly, Gilbert LaFreniere looks to the monastic paradigm as a model for intentional ecocommunities. Such quasi-monastic communities or "ecosteries" would provide a personal and local anchor in a world of uncertainty and cultural transition:

> Such communities, which we might call *ecosteries* on the basis of their debt to ecological principles and the utopian model of the Christian monastery, may furnish the future steady-state society with the same guidance that monasteries of the Dark Ages provided to the rising medieval culture of Western Europe.[63]

Edward Goldsmith, a chief author of *A Blueprint for Survival*, has also drawn a parallel between the decline of the Roman Empire and what he sees as the impending collapse of industrial society, although he argues that in modern times the rate of collapse will be a faster and more cataclysmic event.[64] Goldsmith argues that in both cases the collapse is "the cost of violating in so radical a manner the basic laws of social and ecological organization."[65] In the long term, Goldsmith argues, the family and the community must take over the functions of the State so that social and ecological problems are dealt with at the lowest level possible in Gandhian style "village republics" (on the ground that this would stop such problems being "exported" elsewhere). However, Goldsmith's "self-regulating" communities would

be established "from above" by the State rather than left to emerge organically "from below." In contrast, the other ecoanarchists examined here are consistent in terms of ends and means, preferring exemplary and voluntary local
action rather than state-enforced change.[66]

However, it is Theodore Roszak in his book *Person/Planet* who has presented the fullest defence of ecomonasticism as a solution to the contemporary crisis. Roszak argues that monasticism (he does not confine his attention
to the Benedictine example) provides "a model, a tested, historical paradigm
of creative social disintegration.... where a vital, new sense of human identity and destiny could take root."[67] According to Roszak, this tradition has
shown that it is possible to create relatively self-sufficient and stable domestic economies from very small and humble beginnings. Moreover, it is a tradition that fosters a community that is "simple in means and rich in ends,"
provides an economics of permanence, offers egalitarian fellowship, and is
able to synthesize qualities that have become polarized in modern life such
as the personal/social and the practical/spiritual. In Roszak's view, most
modern political ideologies have overlooked the spiritual and personal
dimension of human experience:

> If the socialist and communist ideologies of our time had not opted
> to become so fanatically antireligious in orientation, they might
> have learned a great truth from the communitarian experience of the
> monasteries. They might have come to see conviviality, not as a dif-
> ficult social duty that must be strenuously inculcated upon us as a
> matter of class consciousness (an approach that only produces mass
> movements), but as a culminating relationship between free and
> unique persons. They might have come to respect the existence of a
> personal reference which supports, but also delimits, the claims of
> the collective will.[68]

In monasteries, in contrast, "the cause of justice can no longer be grounded
in the myth of progress."[69] Rather, economic activity is conducted on a sustainable basis for the satisfaction of basic needs. Moreover, the modern distinction between work and leisure has little relevance in monasteries since
personal fulfilment and a sense of the sacred are found as much *within* work
as outside it. The monastic paradigm, argues Roszak, offers "liberation from
waste and busywork, from excessive appetite and anxious competition that
allows one to get on with the essential business of life, which is to work out
one's salvation with diligence."[70]

Roszak sees the humble beginnings of a modern revival of the ecological community in the decentralist socialist tradition, in Schumacher's economics of permanence, and in the goals of ecological activists and the personal growth movement. He welcomes a proliferation of designs and
experiments for shared and simplified living that are *contemporary* trial-and-

error adaptations of the monastic paradigm, which may hopefully give rise to the "creative disintegration" of modern industrialism. Roszak has no illusions, however, about the likelihood of its immediate or general appeal in modern society. "But then," he observes, "cultural creativity is always the province of minorities."[71]

The ecomonastic paradigm may be seen as providing one kind of holistic solution to the social and ecological problems of the modern world. In particular, it offers a lifestyle that integrates work and leisure, the personal and the political, and the mundane and the sacred. As an ideal, it fulfils the aspirations of ecocentric emancipatory theory insofar as it offers the space for metaphysical reconstruction and cultural, moral, spiritual, political, and ecological renewal. It does this by providing a concrete model of a steady state society that is cognizant of the needs of other life-forms but which nonetheless enables the rounded development of human beings in a cooperative community setting.

However, this model is not without its problems. Indeed, it is instructive to contrast the support for the monastic ideal given by ecocommunal theorists with André Gorz's dismissal of this kind of model—a dismissal that highlights, among other things, the more secular and pragmatic orientation of ecosocialist theorists. Gorz argues that the sanctification of daily activities and the supposed unity of convivial and necessary labor in monastic communities are more apparent than real. In Gorz's view, this is because the realm of "necessary labor" is sublimated by the transformation of external constraints into internal obligations, a sublimation that is mediated by the religious experience.[72] This, argues Gorz, usually takes the form of submission to a spiritual or communal leader who is able "to demand and obtain *submission to necessity as a submission to their own person*. The leader enunciates the law, which is also duty."[73] Gorz argues that "only a dissociation of the spheres of heteronomy and autonomy makes it possible to confine objective necessities and obligations to a clearly circumscribed area, and thus to open up a space for autonomy entirely free of their imperatives."[74] According to Gorz, the only enduring type of commune is that which manages to separate these two spheres, such as the Israeli kibbutz.

It is not necessary to endorse Gorz's own preferred blueprint to acknowledge the well-known dangers that can arise with charismatic leaders, religious or otherwise. Indeed, many supporters of ecomonasticism are also alive to these dangers and are critical of the idea of blind personal allegiance to a particular leader. Roszak, in particular, repeatedly emphasizes the importance of the person, of individual reflection, and of free association as opposed to blind obedience or the surrender of critical judgement.

Nonetheless, Gorz does raise familiar classical arguments in favor of the State as an impartial protector and guarantor of individual liberty—arguments that provide a counterpoint to the ecocommunal reliance on voluntary cooperation and community censure. None of the above ecocommunal theorists satisfactorily address the mechanisms by which recalcitrant community members may be dealt with or the dangers of parochialism and personal affiliation that can often infect the dispensation of "popular" justice. In particular, they have yet to demonstrate how anarchist forms of social control (i.e., noninstitutionalized community censure, coercion, or self-defense) would necessarily be superior to those employed by the modern State (i.e., publicly legitimated institutionalized coercion and protection). Nor do ecocommunal theorists satisfactorily address the question of interregional justice and the problem of disparities in resource endowment *between* local ecocommunities. I shall return to these matters after outlining the distinctive contribution of bioregionalism.

Bioregionalism

Although the exact source of the neologism "bioregionalism" (etymologically, *bioregion* means "life-place") is a matter of some uncertainty, its popularization as a unifying principle celebrating cultural and biological diversity and providing an ecological politics of living-in-place is generally traced to Peter Berg and Raymond Dasmann of the San Francisco Planet Drum Foundation.[75] The term *bioregion,* according to Berg and Dasmann, "refers both to a geographical terrain and a terrain of consciousness—to a place and the ideas that have developed about how to live in that place."[76] Geographically, bioregions are areas having common characteristics of soil, watershed, climate, native plants and nonhuman animals, and common human cultures. Culturally and psychologically, bioregionalism seeks the integration of human communities with the nonhuman world "at the level of the *particular ecosystem* and employs for its cognition a body of metaphors drawn from and structured in relation to that ecosystem."[77] This goal of adapting human communities to the local bioregion is facilitated through the practice of "reinhabitation":

> *Reinhabitation* means learning to live-in-place in an area that has been disrupted and injured through past exploitation. It involves becoming native to a place through becoming aware of the particular ecological relationships that operate within and around it. It means understanding activities and evolving social behavior that will enrich the life of that place, restore its life-supporting systems, and establish an ecologically and socially sustainable pattern of existence within it. Simply stated it involves becoming fully alive in and with a place. It involves applying for membership in a biotic community and ceasing to be its exploiter.[78]

Bioregionalism is principally a North American phenomenon that has grown into a significant tributary of the North American Green movement. In addition to Berg and Dasmann, its principal theorists include Gary Snyder, Kirk-patrick Sale, Ernest Callenbach, David Haenke, Jim Dodge, Morris Berman, and Brian Tokar.[79] Interestingly, too, it enjoys the general support of propo-nents of both social ecology and deep/transpersonal ecology—a point that underscores the broad commonality of these two schools, recent debates notwithstanding.[80] Consistent with the tradition of the ecological community, bioregionalism emphasizes decentralization, human scale communities, cul-tural and biological diversity, self-reliance, cooperation, and community responsibility (both social and biotic). In differs from ecomonasticism mainly in its greater emphasis on protecting, and rehabilitating if necessary, the char-acteristic diversity of *native* ecosystems. This is manifested in its concern to develop a sense of rootedness that is, as Morris Berman put it, "biotic, not merely ethnic."[81] Ecomonasticism, on the other hand, has a more pastoral fla-vor and focuses less on natural history and more on the ideas of personal growth, social cooperation, and spiritual renewal, albeit within an ecological-ly benign setting. These, however, are differences in emphasis only.

To the extent that political forms are discussed by bioregional theorists (many have just been concerned to explore innovative ways of cultivating a bioregional consciousness), the tendency is to promote as a long-term goal a patchwork of anarchist polities linked together by networking and information exchange rather than through a formal state apparatus. In this respect, the gen-eral bioregional response to global problems is encapsulated in the idea of "saving the whole by saving the parts"—the parts, of course, being biore-gions.[82]

In terms of general orientation, bioregionalism is undoubtedly the most ecocentric of all the currents of Green political thought so far examined in part 2. Indeed, as Don Alexander observes, it may be seen as the regional fulfil-ment of Aldo Leopold's land ethic insofar as it sees humanity as a "plain member" rather than a conqueror of the biotic community.[83] This nonanthro-pocentric stance is much more emphatic in bioregionalism than in ecomonasti-cism and social ecology (notwithstanding Bookchin's support of bioregional-ism). However, a case will be made that the political forms suggested by some bioregional theorists are neither the only nor necessarily the best political forms for facilitating the realization of ecocentric goals. The following discus-sion will focus on the work of Kirkpatrick Sale not only because he is a lead-ing bioregional theorist but also because he has had the most to say about political structures. However, it must be borne in mind that much of the biore-gional literature is poetic, inspirational, and visionary and more concerned with cultivating a bioregional consciousness and practice than with presenting a detailed political analysis or developing an alternative national or interna-tional political order. As Peter Berg has explained, bioregionalists may well ally themselves with political movements that have international goals, "but their own primary effort is to solve problems where they live."[84]

Kirkpatrick Sale articulates the view of most bioregionalists in arguing that self-government by the various human communities within a bioregion—possibly linked by a bioregional confederation—offers the best guarantee of social and ecological harmony.[85] (A *confederation* is a mutual association of many autonomous communities or states, each of which retains sovereignty. It should not be confused with a *federation*, which is a mutual association of semiautonomous states or provinces under one, central sovereign state.[86])

On the positive side, the idea that political communities should be based on bioregional contours has much to commend it, particularly in relation to land and water management (indeed, many existing management regimes for internal waters are already modelled along watershed lines). From an educational perspective, bioregionalism plays an invaluable role in underscoring the importance of thinking in terms of ecological relationships, asking where everything comes from and where everything goes, learning to identify, and become "respectful neighbors" with, the local species of flora and fauna. Such an orientation and understanding is crucial to the critical evaluation of existing development decisions. Indeed, some bioregionalists have suggested the formation of "ad hoc watershed shadow governments. Their function would be to serve as moral stewards for specific watersheds and bioregions and to help inhabitants learn the true ecological cost of any proposed development."[87]

At the more practical level, however, bioregionalism is confronted with the problem that linguistic, religious, and cultural boundaries do not necessarily follow bioregional lines. As Don Alexander argues, it is too simplistic to locate human communities on the basis of geography alone.[88] While the patterns of human settlement and movement, and the cultures of many traditional societies, have tended to be influenced quite strongly by geographical criteria, modern transport and communications have meant that regional consciousness in Western society is determined as much by *functional* (and, as already noted, linguistic, religious, and social) criteria rather than formal geographical ones.[89] This seriously challenges the basic assumption on which Sale's entire discussion of political forms is premised, namely, that there is a "clear identity of interest" among the various communities within a bioregion, that they are relatively homogeneous and organically bound together by an ecological identity.[90] What bioregional theorists should ask when considering the issue of institutional design is: what political forms will best promote bioregional goals, given that the many and varied human communities within the many and varied bioregions of the world (however determined—another vexing question) do *not* all possess a bioregional consciousness? Ceding complete political autonomy to the existing local communities that inhabit bioregions will provide no guarantee that development will be ecologically benign or cooperative. Nor will it provide any guarantee that they will form a confederation with neighboring local communities in their bioregion so as to enable proper bioregional management.

Although the goals of bioregional theorists are admirable, their discussion of political forms is based on starting assumptions that are politically naive. (In any event, the flexible notion of bioregions is such that they can vary enormously in size, from the simple ecosystem of a pond to the biosphere as a whole, depending on the scale and criteria used.[91]) Yet any revision of these starting assumptions raises the question as to whether an anarchist polity is indeed the best kind of polity to ensure the realization of bioregional goals, at least in the short and intermediate term. In these and other respects, bioregionalism shares many of the tensions and problems to be found in the general ecoanarchist tradition of which it forms part.

DOES ECOCENTRISM DEMAND ECOANARCHISM?

Despite the important differences between social ecology and ecocommunalism (of which ecomonasticism and bioregionalism are tributaries), both seek to dismantle or bypass the modern nation State and establish decentralized, autonomous, and human-scale communities. For social ecology and bioregionalism, ceding complete political and economic autonomy to such communities is considered the best means of reintegrating human communities into the natural world of which they are part. Ecomonasticism, on the other hand, is less concerned with general political forms and more concerned with planting seeds of cultural renewal through the establishment of exemplary intentional ecocommunities that *coexist* within mainstream society.

However, there are several aspects of ecoanarchism that warrant closer scrutiny. First, how realistic are ecoanarchist assumptions concerning human nature? Second, is decentralization, maximum local democracy, and human scale the only or best means for realizing social and ecological emancipation in the context of the modern world? Finally, how ecologically informed is the ecoanarchist theoretical model of human autonomy? How does it compare to the ecocentric model of autonomy outlined in part 1? Each of these questions will be addressed in turn.

Are Humans "Essentially" Cooperative?

Ecoanarchists argue that decentralization, local democracy, and human-scale institutions maximize opportunities for cooperative self-management. Social ecologists and bioregionalists, in particular, are suspicious of any form of hierarchical social or political arrangement, which they see as thwarting the otherwise spontaneous human impulse to cooperate. In terms of spatial metaphors, they contrast the vertical pyramid with the horizontal web and argue that only the latter is in keeping with the mutualistic nature of ecological relationships. As we have seen, this kind of reasoning is not new to anarchism: indeed, ecoanarchism may be seen as the latest in a long series of attempts by anarchist philosophers to find a model of society in nature.[92] Peter

Kropotkin's *Mutual Aid* is a seminal anarchist tract of this kind insofar as it emphasizes, in seeking to redress the Darwinian stress on competition, the "natural" tendency of animals (including humans) to engage in intraspecies cooperation. Anarchists, as George Woodcock points out, do not all assert that humans are naturally good, but they do fervently assert that humans are naturally social.[93] Roszak goes considerably further in relation to what he calls the mystical anarchist tradition (which includes such figures as Tolstoy, Martin Buber, Walt Whitman, Thoreau, Gustav Landauer, and Paul Goodman):

> Anarchists of this stripe...find their way to a characteristic kind of mysticism, to a warm, intuitive trust in the essential goodness of God, nature and human community. They have known the darkness, but never despair.[94]

Unlike the individualistic liberal model of autonomy, ecoanarchists defend a social model of autonomy that is secured by voluntary cooperation with, and responsibility to, the human scale community of which the individual is part.[95] The dismantling of the State would not lead to social fragmentation, they argue, but rather to spontaneous cooperation and the strengthening of social bonds between people. Antisocial behavior would be dealt with via community censure (as in traditional, small-scale hunting and gathering and horticultural groups) rather than via the abstract and inflexible legal rules laid down by the remote nation State.[96] Moreover, the anarchist assumption that humans are naturally *cooperative,* but are presently corrupted by hierarchical institutions, also stands in contrast to the classical liberal view, which saw humans as naturally self-seeking and in need of restraint through, say, a limited government based on a social contract. It was on the basis of this latter model of human behavior that the survivalist ecopolitical theorists reached a conclusion that is diametrically opposed to that of ecoanarchists: that only a centralized, authoritarian government can rescue us from the ecological crisis and save us, as it were, from ourselves.

Ecoanarchists, in contrast, share a deeply felt desire for humans to cooperate more than they do and a conviction that they can do so in the appropriate social environment. The problem with the ecoanarchist model of human nature, however, is that it conflates people's *potential* nature with their *essential* nature. That is, ecoanarchists present our potential (i.e., better) nature *as* our essential nature and appeal to the reciprocity and mutual aid that they see in nature as evidence that their model of human nature is in alignment with "the natural order of things" and perforce "objectively" right. I have already criticized this ecoanarchist appeal to the natural as adding nothing to the *normative* force of ecopolitical argument and do not intend to repeat these points here. Instead, I want to address the more obvious problems associated with such a model when it comes to rethinking political forms. Specifically, the presumption that humans are "essentially" of a certain nature (i.e., cooperative) and that this nature can be "reawakened" under the right social and institutional circumstances (i.e., anarchism) leads to institutional designs that cannot adequately accommodate human behavior that defies this model of human nature. A more agnostic approach that avoids this kind of essentialism might be to say that the question as to whether humans are "inherently" good or bad (or cooperative or selfish) is meaningless and/or unknowable and to suggest that most, but not all, humans are more or less as "good" as the social and economic institutions of their particular society and culture generally encourages or allows them to be. This acknowledges the importance of social and economic institutions in influencing (as distinct from dictating) human behavior while also allowing for the possibility of idiosyncratic, "antisocial" and, occasionally, pathological behavior. This more agnostic approach thus remains open to the possibility that not everyone will respond to new institutions in the way that their designers might wish—a possibility that is likely to be enhanced if institutional change is rapid and/or pressured by crisis.

Where does this leave ecocentric Green theory? As we saw in chapter 1, all Green theorists reject the authoritarian solution proffered by the survivalists on the grounds that it merely responds to, rather than seeks to challenge and transform, the culture of possessive individualism that characterizes market capitalism. Indeed, Green theorists have been characterized by, inter alia, their concern to find a more lasting solution to the crisis by moving toward a culture of social cooperation and ecological responsibility.

However, ecocentric theorists are confronted with a dilemma created by the urgency of the ecological crisis. While they agree that long term cultural change will provide the most appropriate and lasting solution to the ecological crisis, they recognize that legislation can at least bring about a more immediate response to the crisis. In view of the present rate of global ecological degradation, many threatened habitats, species, and tribal communities are unlikely to survive to see whether mutual aid and ecological restoration will indeed ensue from the ecoanarchists' strategy of withdrawing support from, or seeking to dismantle, hierarchical structures such as the nation State. The ecocentric concern to allow *all* beings (not just humans) to unfold in their own ways demands, at the very least, interim protection of biological and cultural diversity. Such protection can be most effectively achieved by the introduction of State laws and sanctions restraining human conduct— measures that are anathema to ecoanarchists.

More importantly, however, ecoanarchists have not demonstrated that the abolition of, or the withdrawal of support from, "higher" forms of political authority will necessarily lead to *social* emancipation (let alone ecological restoration). Ecosocialism provides a useful counterpoint here in arguing that a democratic State can act as an *enabling* institution that *facilitates* social emancipation by maintaining basic standards of income, health, education, and welfare, and by protecting basic freedoms via the rule of law. In this respect, ecosocialists wish not to abandon but rather to fulfil the promise of parliamentary democracy. Yet Bookchin, for example, insists in his critique of Gorz's ecosocialism, that it must be centralization *or* decentralization, State *or* society, Marxism *or* libertarianism.[97] But is the complete decentralization and devolution of power to the local community the most appropriate means of realizing ecocentric emancipatory goals? Is it not the case that the more a society moves away from the modern welfare State and toward decentralization and diversity, the more we might expect to find disparities in income and social services between communities? How can an anarchist polity mediate between, on the one hand, local interests and, on the other hand, regional and international interests?

The "Other Side" of Decentralization, Local Democracy, and Human Scale

The ecoanarchist case for decentralization, local democracy, and human scale institutions has been heralded by its supporters as the "third" way (i.e., beyond liberal parliamentary democracy and state socialism). Direct democracy at the community level is considered to be essential to the anarchist goals of personal and community empowerment and self-management. As we have seen, ecoanarchists (like all anarchists) argue that the breakdown of the hierarchical nation State and the ceding of power to local communities will enable face to face interaction and direct political involvement in accordance with the ideal of the Athenian polis. Representative democracy, on the other hand, is seen to deliver "top down" decisions that are out of touch with the needs of the local community and therefore less likely to win local allegiance. The complete devolution of power is also seen to liberate what is believed to be a natural human instinct for cooperation as well as make possible self-sufficient and ecologically harmonious local economies.

The problem with this approach is that the general ecoanarchist approach of "leave it all to the locals who are affected" makes sense only when the locals possess an appropriate social and ecological consciousness. It also presumes that local bioregion A is not a matter of concern to people living in bioregion B and that these latter "outsiders" can have no effective input to development decisions made by the inhabitants of bioregion A. Moreover, the rejection of a vertical model of representative democracy in favor of a horizontal model of direct democracy underrates the innovative potential of what might be called the "cosmopolitan urban center" vis-à-vis the "local rural periphery." For example, historically most progressive social and environmental legislative changes—ranging from affirmative action, humans rights protection, and homosexual law reform to the preservation of wilderness areas—have tended to emanate from more cosmopolitan central governments rather than provincial or local decision making bodies.[98] In many instances, such reforms have been carried through by central governments in the face of opposition from the local community or region affected—a situation that has been the hallmark of many environmental battles in the Australian federal system of government. At an even "higher" level, bodies such as the International Court of Justice and the World Heritage Committee are salutary reminders of the ways in which institutions created by international treaties can serve to protect both human rights and threatened species and ecosystems from the "excesses" of *local* political elites. Indeed, there is a large number of Green social and environmental reforms, ranging from the redistribution of resources from rich to poor countries to the abatement of the greenhouse effect, that can be effectively implemented *only* via international agreement between existing nation States. Successful ecodiplomacy of this kind is more likely to be achieved by the retention and reform of a democratically accountable State that can legitimately claim to represent in the international arena at least a majority of people in a nation. While unilateral action by "right minded" citizens in local bioregions is to be encouraged, it will have minimal effect for as long as recalcitrant neighboring local communities and regions continue to "externalize" their environmental costs. In view of the urgency and ubiquity of the ecological crisis, ultimately only a supraregional perspective and multilateral action by nation States can bring about the kind of dramatic changes necessary to save the "global commons" in the short and medium term. It must be emphasized that none of these arguments are intended to deny the innovative potential of local and municipal action and the importance of enhancing local autonomy, nor the many obstacles facing international agreement. I am merely concerned to point out the *two-edged* nature of the argument for the complete devolution of political and legal power to local assemblies. We need not only to act locally and think globally, but also to *act* globally.

Decentralization can also be ecologically and socially problematic when pressed too far. Indeed, there are strong ecological arguments *against* complete decentralization in terms of a more uniform distribution of human settlement. As Robert Paehlke has pointed out, urban settlements are a less ecologically stressful and more energy efficient way of accommodating large numbers of people on the land than dispersing the human population more thinly and widely throughout existing wilderness and rural areas.[99] Cities also provide a cosmopolitan culture in contrast to the parochialism that can often be found in small rural communities. In criticizing the rural romantic current in the Green movement, Stephen Rainbow has observed that

> while the urbanization accompanying industrialization has many negative side-effects it has also bred many social-emancipatory movements and facilitated a much more interesting life for many of the city's inhabitants than they might have enjoyed in the countryside. A simple rural romanticism is far too hard to sustain against

the reality of history, and it is not the only basis upon which the Green desire for community can be built.[100]

While there are some strands of ecocommunalism that are vulnerable to these criticisms and while many ecoanarchists (for example, Roszak) are critical of *extensive* urbanization and *large* cities, most ecoanarchists recognize the ecological *and* cultural advantages of cities.[101] Indeed, far from advocating the demise of cities as cultural and civic centers, bioregionalists and social ecologists advocate reinvigorating and greening cities and developing an urban "life-place" consciousness. Peter Berg, for example, has developed a proposal for Green cities, while Bookchin advocates the reinvigoration of city neighborhood assemblies in his "New Municipal Agenda."[102] An ecocentric perspective is more consistent with a diversity of human dwelling patterns (that is, medium sized cities *and* decentralized communities) than with all of one or the other. This would be in a context of a lower human population and with less giant agglomerations of human settlement.

Insisting too emphatically on decentralization, local political autonomy, and direct democracy can also compromise the ecocentric goal of social justice. I have already noted that the more we move away from the modern welfare State to local autonomy, the less we can expect to find the same levels of wealth, welfare, and social services among different local communities. This is because there is no longer an effective central decision-making forum able to redistribute resources between regions to overcome, say, interregional disparities in resource endowment, wealth, and public infrastructure or provide relief in times of hardship or disaster.[103] This has particular relevance to developing countries. Bioregionalists, for example, do not address the fact that not all bioregions are equally endowed with the resources that enable the satisfaction of basic human needs. To what extent do we allow migration, trade, or compensation to promote social equality on an inter-bioregional and international basis? Won't migration be likely to disrupt the bioregional goals of social cohesion and self-reliance and put a strain on the carrying capacity of the well-endowed bioregions? Can we afford to depend on the ecoanarchist reliance on goodwill and voluntary networking to resolve these many tensions? In view of the centrality to ecocentrism of promoting a level of resource consumption that can be sustained for *all* humans and that is compatible with the flourishing of nonhuman communities, the case for an "enabling State" to facilitate this transition begins to appear much more robust than the ecoanarchist case for spontaneous cooperation and voluntary interregional networking.

An ecocentric perspective would seem to be more consonant with a political decision making framework that can represent, address, and resolve—or at least accommodate—social and cultural differences both within and across communities and regions. This would require a multitiered institutional framework that ensures the dispersal of political power *between* the center and periphery (rather than the concentration of political power in either the center or periphery) in order to provide checks and balances in both directions. As we shall see, such a multitiered democratic political structure is also more compatible with an ecocentric worldview than the simple web-like, horizontal structure of ecoanarchism, insofar as it recognizes the layered interrelationships between parts (social and ecological) of larger wholes.

THE ECOANARCHIST MODEL OF AUTONOMY AS SELF-MANAGEMENT

The ecoanarchist model of autonomy is encapsulated in the notion of political, economic, and social self-management. Ecoanarchist theorists generally see any kind of external or "other-directedness," as distinct from internal or "self-directedness," as encroaching on this fundamental norm. As we have seen, the kind of "self" that is to be "managed" in the ecoanarchist model of autonomy-as-self-management is a cooperative, social self rather than the atomistic and self-seeking individual of classical liberalism. Selfhood, according to social ecology, means having the power, competence, and necessary social development to be a fully participating citizen in the polis; until such selves are minimally attained, argues Bookchin, self-management becomes a contradiction in terms.[104] If ecosocialism may be seen as pressing forward the liberal notion of autonomy by insisting on both political *and* economic democracy (both representative and direct), then ecoanarchism may be seen as pressing forward this notion yet further by insisting on *direct* democracy in *all spheres of life*. Unlike ecosocialists, most ecoanarchists do not accept representative democracy or the delegation of political or economic power to any higher authority. While some ecoanarchists accept the idea of larger coordinating political units beyond the local community, such as confederations of self-managing communities, these larger units are to remain thoroughly subservient to the member communities. As Bookchin explains, "the living cell which forms the basic unit of political life is the municipality from which everything else must emerge: confederation, interdependence, citizenship, and freedom."[105]

How ecologically informed is this ecoanarchist model of autonomy? Is it compatible with the process-oriented, ecological model of internal relations that informs ecocentrism? As we saw in chapters 2 and 3, this ecological model posits the individual human as one kind of relatively autonomous organism that is, like other entities, embedded in a myriad of multilevelled relationships of many different kinds (in the case of humans, this includes physiological, psychological, social, and ecological relationships). Moreover, these many relationships are intrinsically dynamic, nonlinear, and flexible in responding to feedback (within certain parameters)—often in unpredictable ways. In the light of this picture of ecological reality, we may imagine ourselves as cells in the body of Gaia, forming part of larger wholes that themselves have unique properties and which form part of still larger wholes (ecosystems, ethnic and cultural groupings, states, geopolitical regions, and so on).

From the perspective of this dynamical ecological model, neither the direct democracy of an anarchist polity nor a completely centralized bureaucratic State possesses the flexibility and resilience to deal with international pressures or local needs respectively.[106] The simple web-like, horizontal decision making structure of ecoanarchism, while flexible and responsive to the needs of local communities, has no *built-in* recognition of the "self-management" interests of similar or larger social and ecological systems that lie beyond the local community. This is because a confederal body cannot proceed without the voluntary cooperation of its member units and cannot override the decisions of its member units; the latter are determining, but not determined. The ecoanarchist insistence on the sovereignty of the local community can therefore admit only of differentiation of function and competence (and hence sovereignty) by individuals and groups *within* a local community, not beyond it. Of course, many ecoanarchists (particularly bioregionalists) subscribe to the ecological model of internal relations and seek the cultivation of an ecological consciousness; however, this recognition is not adequately reflected in the organizational forms recommended by ecoanarchists.

Conversely, the centralized bureaucratic State can purport to represent the interests of a larger social and ecological whole but its "top heavy" and "top down" hierarchical structure cannot respond flexibly to the special needs and interests of local communities. Moreover, such a concentration of power has the potential to lead to far-reaching domination.

What is the most appropriate locus of self-management in the context of an interconnected world made up of many different kinds of relatively autonomous autopoietic entities? Clearly, whatever social and ecological "whole" we identify as the locus of political self-management will always be partial insofar as there will always be other kinds of social and ecological "wholes" or communities with their own, somewhat different "interests" to be represented. This suggests the need for a layered and flexible decision-making structure—with a two-way flow of information—to ensure that political and economic power is not excessively concentrated at any one level, whether "top" *or* "bottom." It means a greater dispersal of power both "down" and "up" in the sense of a simultaneous devolution of some areas of legal and fiscal power from the nation State to local communities as well as the ceding of other areas of legal and fiscal power from the nation State to international democratic forums. At present, the concentration of political power in the nation State not only gives rise to inflexibility toward the special needs of local communities but also gives rise to a limited, nationalistic notion of security that hampers international efforts (urged by many non-government organizations) toward ecological cooperation and a more equitable distribution of resources between developed and developing nations. Indeed, it is already possible to discern—particularly in Europe—a lessening of state power and national sovereignty brought about by the strengthening of local democracy, on the one hand, and greater regional and international cooperation and agreement brought about by the recognition of mutual ecological interests, on the other hand.

In terms of economic management, this model is inconsistent with extensive State economic planning or a complete "command economy" since it seeks the dispersal of both political *and* economic power in both State and private hands. It is, however, consistent with government intervention in the economy (whether local, provincial, or national) to break down excessive concentrations of market power or to ensure that the operation of the market does not compromise ecological integrity or social justice. Finally, such a model is consistent with a greater degree of local community ownership and control of the means of production as well as cooperative enterprises and worker self-management.

In this chapter I have argued that a multitiered, democratic decision making framework is more consistent (both theoretically and practically) with an ecocentric perspective than the political forms advocated by ecoanarchism. Of course, a democratic decision-making framework does not in itself guarantee social and ecological *outcomes* that are consistent with an ecocentric perspective. In this respect, the ecological and cultural renewal envisioned by ecoanarchists, such as cultivating a bioregional consciousness, social responsibility, and spirit of civic participation, provide the crucial lifeblood of a successful ecocentric polity. Nonetheless, I have sought to show that a democratically accountable nation State (operating in the context of a relatively decentralized, multitiered governmental framework) is much better placed than a large number of autonomous local governments when it comes to providing ecodiplomacy, interregional and international redistributive justice, and the protection of uniform human rights and freedoms via the rule of law. It is also more compatible with an ecocentric worldview than the simple web-like, horizontal structure of ecoanarchism, insofar as it recognizes the layered interrelationships between parts (social and ecological) of larger wholes. While it gives prima facie priority to the needs of local communities, it recognizes that the maintenance of "healthy" and diverse ecosystems and bioregions is a matter of concern not only to the people who inhabit them.

CONCLUSION

We saw in chapter 1 that the ecological crisis has been analyzed by political theorists as reflecting a crisis of participation, a crisis of survival, and a crisis of culture and character. I have argued that Green political theory may be characterized by its concern to reconcile and solve all three of these interrelated crises while also offering new opportunities for social emancipation, cultural renewal, and an improved quality of life. However, there are many different ecophilosophical orientations that jostle together under this very broad Green umbrella. I have argued that only a thoroughgoing ecocentric Green political theory is capable of providing the kind of comprehensive framework we need to usher in a lasting resolution to the ecological crisis. When compared to the other ecophilosophical approaches examined in this inquiry, an ecocentric approach has been shown to be more consistent with ecological reality, more likely to lead us toward psychological maturity, and more likely to allow the greatest diversity of beings (human and nonhuman) to unfold in their own ways. Indeed, ecocentrism may be seen as representing the cumulative wisdom of the various currents of modern environmental thought. Moreover, I have shown that an ecocentric perspective has already been considerably developed—at least in an ethical, philosophical, and psychological direction—in autopoietic intrinsic value theory, transpersonal ecology, and ecofeminist theory.

Whereas part 1 was mainly concerned to trace, explain, and defend the philosophical foundations of ecocentrism, part 2 was mainly concerned to find out how, through a critical dialogue with the major streams of Green political thought, an ecocentric philosophical perspective might be fleshed out in a political and economic direction. This inquiry also entailed a critical examination of the ecophilosophical orientation of each of these new traditions of ecopolitical thought.

In terms of overall ecophilosophical orientation, ecoanarchism proved to be the most ecocentric of the Green political theories examined. In contrast, the three families of socialism (eco-Marxism, Critical Theory, and ecosocialism) all belonged to the anthropocentric stream of Green political thought. However, the degree of anthropocentrism tended to diminish as we moved from orthodox eco-Marxism to humanist eco-Marxism (including Critical Theory) and then to ecosocialism.

179

Indeed, it is illuminating to compare the different forms in which anthropocentrism has been characteristically expressed in these three families of socialist thought. As we saw in chapter 3, social and political theories are often legitimated on the grounds that they will enable the realization of whatever are taken to be the most important qualities of "humanness." Whatever these qualities, they are defended as providing us with our human dignity or moral worthiness in that the cultivation of these qualities helps to distinguish us from, and "raise" us above, the rest of nature. For example, Andrew McLaughlin points out that if we are considered to be essentially *homo faber,* the fabricator or tool maker,

> then our history appears as a nearly linear progression forward in the domination of nature through the development of increasingly powerful technologies.... These advances appear as a progressive realization of human nature and yield a clear definition of human progress.[1]

Alternatively, if the criterion is taken to be our unique ability to communicate via language, then the more we can perfect that communicative ability by, say, aspiring toward the establishment of an "ideal speech situation," then the more we may be said to have progressed *qua humans.* More generally, if we are taken to be free, creative beings of *praxis* who consciously transform the world, ourselves, and our social relations, then the removal of any impediments to, or the creation of conditions that enhance, free self-determining activity represents progress in terms of human self-realization.

Each of these examples—taken from Marx, Habermas, and democratic socialist thought in general—indicates how particular conceptions of what is special about being human can determine what constitutes human progress or human self-realization. Where socialists differ from liberal political theorists is in their more concerted attempt to create the social, political, and economic conditions that will enable *all* members of the human family to realize whatever is taken to be their worthy and special human potential. As we have seen, not all of these conceptualizations of what is special about being human are equally problematic from an ecocentric perspective. The orthodox eco-Marxist approach turned out to be the most *active* kind of discrimination against the nonhuman world. Indeed, the orthodox eco-Marxist conception of human beings as *homo faber* amounted to a positive invitation to expand the "forces of production" and subdue the rest of nature to the greatest possible extent so that humans could control the nonhuman world for human purposes. By contrast, the humanist eco-Marxist and ecosocialist conceptualizations of what it means to be human were considerably more tempered and may be seen as constituting passive, as distinct from aggressive, expressions of anthropocentrism insofar as they discriminate against the nonhuman world, in varying degrees, generally by omission rather than commission.[2] That is, while they do not actively promote the domination of the nonhuman world, their exclusive focus on human well-being means that they systematically fail to consider the special interests and needs of the nonhuman world.

In the case of humanist eco-Marxism, the price of overcoming human alienation from "inner" and "outer" nature is the thoroughgoing domestication of the nonhuman world in the name of human self-expression. Habermasian Critical Theory, in contrast, contains both passive and aggressive expressions of anthropocentrism. Habermas's theory of communicative ethics, for example, is passively anthropocentric, in that it is restricted to serving the interests of human speaking participants. Here, the dignity and rights of the human subject are secured, while the nonhuman world is simply neglected. However, Habermas adopts a more explicit anthropocentric posture in his insistence that science can know nature only in instrumental terms. As we saw in chapter 5, this limited conceptualization of science serves to actively reinforce rather than challenge the domination of the nonhuman world.

Ecosocialism signals a partial return to the early Frankfurt School's critique of instrumental reason. However, this critique ultimately comes to rest on the argument that it is wrong to dominate nonhuman nature *because* it leads to the domination of people. In other words, the ecological ethic advanced by ecosocialism shows no direct concern for the welfare of nonhuman nature. This is reflected in the ecosocialist *analysis* of the environmental crisis, and in the ecosocialist interpretation of the message of ecology (i.e., that it can only inform us of the limits of human activity and cannot inspire new political values). Notwithstanding this passive anthropocentrism, I suggested that ecosocialism has the potential to be revised in an ecocentric direction simply by extending its theoretical horizons, that is, by extending its fundamental norm of respect for all persons to encompass respect for all life-forms and ecological entities. This would bring about the necessary shift in the ecosocialist response to wilderness and human population issues while incorporating other worthy ecosocialist goals such as the new internationalism, production for human need, and economic democracy.

However, it is one thing to have an appropriate ecocentric philosophical orientation and another thing to find the political and economic institutions that will best realize ecocentric goals. Here we found that none of the political and economic reforms offered by the various Green political theories examined were likely to realize ecocentric objectives in the foreseeable future in the absence of extensive revision.

Although ecoanarchism proved to be the most ecocentric of the various emancipatory theories examined in terms of its general sensibility, this ecocentric sensibility was not adequately reflected in the antistatist political forms recommended by ecoanarchists. However, before drawing together this particular criticism of ecoanarchism, it must be pointed out that many of the ecoanarchist theories examined in chapter 7 have not even addressed, or address in only a cursory way, the question of institutional design at a general societal level. In particular, many ecoanarchists are less concerned with bringing about general economic and political structural change (at least in the first instance) and more concerned to facilitate cultural renewal by establishing exemplary ecological communities within the "shell" of existing society. By encouraging people to exercise existing political and economic freedoms and engage in small scale, local experiments of this kind, these ecoanarchists are concerned to develop peaceful ways of facilitating what Roszak has called the "creative disintegration" of industrial society. In this respect, ecoanarchism has made, and will continue to make, a vital contribution to the development of an appropriate ecocentric emancipatory *culture*— a culture that is infused with a democratic ethos and a sense of personal, civic, and ecological responsibility. As Alex Comfort has observed, anarchism must be seen more as an attitude than as a program:

> Anarchists do not plan revolutions—but when they become numerous,...[anarchist] thinkers constitute active, unbiddable and exemplary lumps in the general porridge of society. If numerous enough, they begin to affect the types of choices which societies make.[3]

However, when we turn to those ecoanarchists who have addressed the matter of political organization at the general societal level, we found that an antistatist framework did not prove to be the best means of serving ecocentric goals. Indeed, we saw that there are sound ecocentric reasons for not ceding complete political power to small, local communities. This is because the ecoanarchist defence of local sovereignty provided no firm institutional recognition of the many different layers of social and ecological community that cohere *beyond* the level of the local community. Moreover, the ecoanarchist case for ceding sovereignty to local communities rested on a somewhat naive and overoptimistic model of human nature (i.e., it was assumed/expected that local communities will *voluntarily* consider and respect the interests of other local communities). "Small" is not "beautiful" when the rule of the nation State is replaced with the rule of the local community in circumstances where that local community is impoverished and its local ecosystem is poorly endowed or denuded, or where the local community chooses, or is forced by economic necessity to adopt, a development path that undermines the local ecosystem. We face a highly unstable future, and we cannot afford to relinquish the institutional gains of parliamentary democracy and the (however imperfect) checks and balances they provide against the abuse of power—at least not until such time as an ecocentric consciousness has substantially permeated our political culture. Instead, we should be concerned to *revitalize* these institutional gains by strengthening such checks and balances. This will enable the forum of parliament to be used to democratize society at large by gradually breaking down excessive concentrations of political and economic power. This includes breaking down the sovereignty of the nation State and dispersing appropriate areas of political power both "up" (i.e., to interregional and international democratic decision making bodies) and "down" (i.e., to local decision making bodies such as municipal governments). A multilevelled decision making structure of this kind is more theoretically compatible with an ecocentric perspective than the kind of complete local sovereignty defended by ecoanarchism, because it provides a far greater institutional recognition of the different levels of social and ecological community in the world. Such a multilevelled decision-making structure is also better able to implement ecocentric emancipatory goals. In particular, it is better able to secure the international, interregional, and intercommunity agreement that is essential to dealing with the ecological crisis and better placed to secure the maintenance of basic standards of income, health, education, and welfare between communities, regions, and nations.

The ecosocialist approach to strategy, institutional design, and the role of the State provides an important counterpoint to the ecoanarchist response to these issues. Indeed, this tension is currently being played out in day-to-day Green politics between the realist and the fundamentalist wings of the Green movement and Green parties, that is, between those who want to take the electoral route and gain political power and those who want to bring about change at the grassroots level and thereby avoid being corrupted by what is seen as the "power politics" of hierarchical institutions.[4] As we have seen, the prime concern of most ecoanarchists is with "right action" and authenticity rather than with political expediency or the conquest of power. The political goal is generally one of inducing or inviting change through education and exemplary action, leading to the withering of support for hierarchical institutions. If this proves ultimately to be ineffectual, then so be it, since ecoanarchists—consistent with their high ideals of individual and community autonomy and personal responsibility—seek to avoid deciding a way of life for others.

Ecosocialists, however, are critical of ecoanarchism for being "voluntarist," utopian, and ultimately marginal and ineffectual. Ecosocialists are concerned with the *practical* negation of the economic and political domination that arises under capitalism and totalitarianism and, to this end, are concerned to develop a theoretical understanding of society that will generate a successful politics of transition on a far-reaching scale. In this respect, ecosocialists argue that it is not enough simply to encourage people to engage in voluntary, exemplary action by means of an appeal to the common good—as if everyone has an equal interest in social and ecological reform. Rather, they argue that ecoanarchism must recognize and address the fact that certain classes have a material interest in maintaining the status quo. Ecosocialists also argue that it is utopian and idealistic to encourage only voluntary action and "small experiments" and not address *systemic* problems. In particular, they argue that it is naive to bypass the State in the expectation that it will wither away without a struggle by the simple transfer of allegiances. Finally, they argue that ecoanarchists either ignore or do not adequately address questions such as uniform civil rights, basic welfare and redistributive justice, revenue raising, intergovernmental cooperation, and international representation—a case that I have thoroughly endorsed.

Indeed, most ecosocialists regard the State—the agent of the "collective will"—as playing a vital role in controlling the operation of market forces and in laying down the framework for a socially just and ecologically sustainable society. This also includes breaking down concentrations of political and economic power in large corporations. Yet we saw that the ecosocialist acknowledgement of the need for a more equitable dispersal of power is somewhat undercut by the degree to which most ecosocialists are prepared to augment the economic powers of the State through the establishment of new public enterprises and new economic planning (as distinct from just regulatory) endeavours designed to replace the market economy. Although ecosocialists argue that the potential for bureaucratic domination would be offset by enhanced opportunities for decentralized democratic participation, it is hard to see how economic planners in an ecosocialist society would succeed in effectively implementing a series of coherent economic plans without a central coordinating unit with considerable political and economic power. In short, ecosocialists have yet to resolve with the tension between their quest for self-management and participatory democracy and their reliance on centralized institutions to carry out far-reaching social and economic reforms.

Although ecosocialists have accurately highlighted many of the contradictions between market rationality, on the one hand, and social justice and ecological sustainability, on the other hand, I have argued that it does not necessarily follow that the drastic contraction of the market economy is the only or best means of addressing these contradictions. This is because such a step introduces many new problems of a different order (e.g., black markets and enhanced bureaucratic domination, to name the most familiar). I have argued that an alternative and preferable solution is to retain the market as the basic system of resource allocation in most areas (and here there should be plenty of room to argue for exceptions) but to introduce a greater range of macro-controls on market activity in order to break down excessive concentrations of economic power, protect ecosystem integrity and biological diversity, and promote greater equality of social opportunity. Such macro-controls would serve to discipline entrepreneurial activity and channel it into those areas that produce no adverse social and ecological consequences. A more heavily circumscribed market economy that is scaled down in terms of material-energy throughput would still preserve competition between firms, and investment would still be guided by consumer spending (although it would need to be much more ecologically and socially informed spending, aided by enhanced consumer education).

Having drawn together the main insights from our critical dialogue with the major currents of Green political thought, we can now sketch a broad-brush picture of what an ecocentric polity might look like. Such an ecocentric polity would be one in which there is a democratic state legislature (which is part of a multilevelled decision-making structure that makes it less powerful than the existing nation State and more responsive to the political determinations of local, regional, and international democratic decision-making bodies); a greater dispersal of political and economic power both within and between local communities; a greater sharing of wealth both within and between local communities; a far more extensive range of macro-controls on market activity; and the flowering of an ecocentric emancipatory culture. Of course, this picture merely represents the very broad outlines of an ecocentric polity. There remains plenty of room for considerable diversity in terms of institutional and economic design. Moreover, although the flowering of an ecocentric emancipatory culture is a necessary part of an ecocentric polity, the proposed general institutional and economic framework outlined in chapters 6 and 7 has been defended as being the most conducive to steering our way out of the ecological crisis in the present context *irrespective* of whether an ecocentric perspective is widely shared. Unlike bioregionalists and many other ecoanarchists, I do not assume that handing over more power to local communities will necessarily make them Green, like-minded and "good." Higher order legislative assemblies, institutional checks and balances, the protection of basic political freedoms (e.g., of speech and assembly), and the rule of law are essential to prevent excessive parochialism and the abuse of power.

Nonetheless, I have argued that the cultivation of an ecocentric culture is crucial to achieving a *lasting* solution to the ecological crisis. This is because it is only in those political communities in which an ecocentric sensibility is widely shared that there will be a general consensus in favor of the kinds of far-reaching, substantive reforms that will protect biological diversity and life-support systems. At a minimum, those who share an ecocentric perspective understand and accept the need for the protection of large tracts of representative ecosystems; the development of a humane population policy that respects the carrying capacity of ecosystems and the "rights" of other species to share the Earth's life-support system; a fundamental reevaluation of human needs, technologies, and lifestyles in such a way as to minimize energy and resource consumption and minimize or eliminate pollution; and the provision of adequate compensation whenever ecological reforms are likely to produce inequitable consequences for certain social groups, classes, or nations.

However, it is also necessary to acknowledge that for those who do not share an ecocentric perspective (and this includes many people who declare themselves to be environmentalists), many of these changes appear threaten-ing, bad for business, unnecessary, or at the very least considerably prema-ture until such time as we have more evidence of impending ecological dis-aster. One of the urgent tasks of ecocentric philosophers, political theorists, politicians, and activists is not only to ensure that there is a regular and accu-rate flow of information to the general public on ecological issues but also to stimulate a far more extensive political debate on environmental values. This includes calling into question long-standing and deeply held anthropocentric assumptions and prejudices. It also requires the generation of new ways of seeing and new visions of an alternative ecological society that enable people to imagine or visualize what it might be like to live differently, and with greater ecological security.

One way in which emancipatory writers have contributed to new ways of seeing is through the utopian novel or political tract. In an illuminating discussion on utopianism, Ruth Levitas distinguishes between utopianism as form (i.e., the Blueprint) and utopianism as function (i.e., venturing beyond the given, releasing the imaginative faculties, and providing a heuristic of future possibilities): "The point then becomes not whether one agrees or dis-agrees with the institutional arrangement, but rather that the utopian experi-ment disrupts the taken-for-granted nature of the present."[5] Ecoanarchism, along with the new genre of "ecotopian" literature by ecoanarchist and ecofeminist writers, provides an invaluable service to Green political dis-course by stimulating our imaginative faculties and providing what E. P. Thompson has called "the education of desire"—an education that "opens the way to aspiration."[6] To dismiss this educative function of utopianism is to deny a major impulse to progressive political engagement and severely limit the means by which we may examine what passes for "common sense."

Nonetheless, the Green movement will ultimately stand or fall on its ability to generate *practical* alternatives to the advanced industrial way of life. As ecosocialists such as E. P. Thompson and Raymond Williams have rightly argued, it is important to connect utopian aspirations with analysis and human experience rather than allow such aspirations to settle as a mere mental compensation for, or a means of escape from, the shortcomings of the status quo. To be realized, the aspirations released by utopianism must be critically related to one's knowledge of the present, thereby uniting desire with analysis and leading to informed cultural, social, and political engage-ment. The ecocentric Green movement needs idealists *and* pragmatists, cre-ativity *and* critical analysis, grassroots activity *and* institutional support if it is to achieve its long-term aims.

Documentation

NOTES TO INTRODUCTION

1. Lynn White, Jr., "The Historical Roots of Our Ecologic Crisis, *Science* 155 (1967): 1204.

2. Steven Lukes, *Individualism* (New York: Harper and Row, 1973), 49.

3. H. J. Blackham, *Humanism* (New York: International Publications Service, 1976), 102.

4. David Ehrenfeld, *The Arrogance of Humanism* (Oxford; Oxford University Press, 1981), xiii–xiv.

5. Throughout this inquiry I use the term *nature* to encompass both the human and the nonhuman worlds and avoid the juxtaposition "human vs. nature" (which misleadingly suggests that humans are not part of nature).

6. White, "The Historical Roots of Our Ecologic Crisis," 1206.

7. Alvin Gouldner, *The Coming Crisis of Western Sociology* (London: Heinemann, 1973), 34.

NOTES TO CHAPTER 1

1. For an overview of Green parties around the world, see Sara Parkin, *Green Parties: An International Guide* (London: Heretic, 1989).

2. The birth of the modern environmental movement is typically associated with the publication of Rachel Carson's international best seller *Silent Spring* in 1962 (New York: Fawcett Crest, 1962. Reprint. Harmondsworth, U.K.: Penguin, 1970).

3. For a general overview, see John McCormick, *The Global Environmental Movement: Reclaiming Paradise* (London: Belhaven, 1989), chapter 3 ("The Environmental Revolution [1962–1970]").

4. John Rodman, "Paradigm Change in Political Science: An Ecological Perspective," *American Behavioral Scientist* 24 (1980): 65.

5. Ibid.

6. Carson, *Silent Spring*; Murray Bookchin [Lewis Herber, pseud.], *Our Synthetic Environment* (New York: Knopf, 1962); and Charles Reich, *The Greening of America* (Harmondsworth, U.K.: Penguin, 1971).

7. Hugh Stretton, *Capitalism, Socialism and the Environment* (Cambridge: Cambridge University Press, 1976), 1.

8. For an influential critique of this kind, see Hans Magnus Enzensberger, "A Critique of Political Ecology," *New Left Review* 84 (1974): 3–31. For a reply, see Robyn Eckersley, "The Environment Movement as Middle Class Elitism: A Critical Analysis," *Regional Journal of Social Issues* 18 (1986): 24–36 and Riley E. Dunlap and Denton E. Morrison, "Environmentalism and Elitism: A Conceptual and Empirical Analysis," *Environmental Management* 10 (1986): 581–89.

9. See, for example, Ronald Inglehart, *The Silent Revolution: Changing Values and Political Styles Among Western Publics* (Princeton, N.J.: Princeton University Press, 1977) and Ronald Inglehart, "Post-Materialism in an Environment of Insecurity," *The American Political Science Review* 75 (1981): 880–900. For a critical evaluation of the major sociological explanations (including Inglehart's) put forward to account for the predominant New Class composition of the ecology movement and broader Green movement, see Robyn Eckersley, "Green Politics and the New Class: Selfishness or Virtue?" *Political Studies* 37 (1989): 205–23.

10. Herbert Marcuse, *One Dimensional Man* (London: Routledge and Kegan Paul, 1964. Reprint. London: Abacus, 1972) and Jürgen Habermas, *Toward a Rational Society: Student Protest, Science, and Politics*, trans. Jeremy J. Shapiro (London: Heinemann Educational Books, 1971).

11. See, for example, Bookchin [Lewis Herber, pseud.], *Our Synthetic Environment*; Bookchin, *Post-Scarcity Anarchism* (Berkeley: Ramparts, 1971); Theodore Roszak, *The Making of a Counterculture: Reflections on the Technocratic Society and its Youthful Opposition* (London: Faber & Faber, 1970, reprint ed., 1973); Roszak, *Where the Wasteland Ends: Politics and Transcendence in Postindustrial Society* (Garden City, N.Y.: Anchor/Doubleday; reprint ed., London: Faber and Faber, 1973); and Reich, *The Greening of America*.

12. George Katsiaficas, *The Imagination of the New Left: A Global Analysis of 1968* (Boston, Mass.: South End, 1987), 5.

13. See William Leiss, *The Domination of Nature* (Boston: Beacon, 1974).

14. The other two pillars are ecology and non-violence. See, for example, *Die Grünen, Programme of the German Green Party* (London: Heretic Books, 1983), 7–9.

15. Donella H. Meadows, Dennis L. Meadows, Jorgen Randers, and William W. Behrens III., *The Limits to Growth: A Report for the Club of Rome's Project on the Predicament of Mankind* (New York: Universe, 1972) and Edward Goldsmith et al., *Blueprint for Survival* (Boston: Houghton Mifflin, 1972; reprint ed., Harmondsworth, U.K.: Penguin, 1972).

16. Earlier warnings can be found in Harrison Brown, *The Challenge of Man's Future* (New York: Viking, 1954); Bookchin, *Our Synthetic Environment* (1962); Carson, *Silent Spring* (1962); Stuart L. Udall, *The Quiet Crisis* (New York: Holt, Rinehart, and Winston, 1963); and Paul Ehrlich, *The Population Bomb*, rev. ed., (London: Pan/Ballantine, 1972). Other warnings in the early 1970s include Paul Ehrlich and Anne Ehrlich, *Population, Resources, Environment* (San Francisco: Freeman, 1970); Richard A. Falk, *This Endangered Planet: Prospects and Proposals for Human Survival* (New York: Vintage, 1972); and Barry Commoner, *The Closing Circle: Nature, Man and Technology* (New York: Knopf, 1971; reprint ed., New York: Bantam, 1972).

17. *The Ecologist* magazine's "blueprint" proposed a quite radical and specific set of measures to deal with the ecological crisis and advocated the need for a decentralized, steady state society. This went much further than the Club of Rome's call for more research and for concerted national and international action to attain a state of "global equilibrium." See the commentary by the Executive Committee of the Club of Rome, in Meadows et al., *The Limits to Growth*, 185–97.

18. Joseph Campbell, *Myths to Live By* (New York: Viking, 1972; reprint ed. New York: Bantam, 1973), quoted by Yaakov Jerome Garb in "The Use and Misuse of the Whole Earth Image," *Whole Earth Review*, March 1985, 18.

19. The aims of the British party were "to create a self-reliant, community based way of life within the framework of a stable economy and a just, democratic society, so that people may live in harmony with each other and the rest of the natural environment by acknowledging and adapting to the limitations of the earth's finite resources." See Alistair McCulloch, "The Ecology Party and Constituency Politics: The Anatomy of a Grassroots Party," Paper presented at the Annual Conference of the Political Studies Association, University of Newcastle-upon-Tyne, April 1983. The quotation is taken from an Ecology Party pamphlet, *The Politics of Ecology* (London: n.p., 1979), 4. For a more recent discussion of the changing fortunes of this party, see Jonathon Porritt and David Winner, *The Coming of the Greens* (London: Fontana, 1988), 60–62 and 76–78.

20. For a methodological critique, see H. S. D. Cole, C. Freeman, M. Jahoda and K. L. R. Pavitt, *Thinking About the Future: A Critique of the Limits to Growth* (London: Chatto and Windus, 1973). For an optimistic alternative, see Herman Kahn, William Brown, and Leon Martel, *The Next 200 Years* (London: Associated Business Programmes, 1976; reprint ed., London: Abacus, 1978). For a general discussion of the debate, see John Gribbon, *Future Worlds* (London: Abacus, 1979), chapter 1 ("Boom or Gloom? The Great Debate").

21. Mihajlo Mesarovic and Eduard Pestel, *Mankind at the Turning Point* (New York: Dutton, 1974), 142.

22. See, for example, Enzensberger, "A Critique of Political Ecology."

23. Gerald O. Barney, study director, *The Global 2000 Report to the President: Entering the Twenty-First Century*, vol. 1. (Harmondsworth, U.K.: Penguin, 1982), 1.

24. Lester Brown, gen. ed., *State of the World 1984* (New York: Norton, 1984) and annually thereafter; World Commission on Environment and Development (Chaired by Gro Harlem Brundtland), *Our Common Future* (Oxford: Oxford University Press, 1987). See also Richard J. Barnet, *The Lean Years: Politics in the Age of Scarcity* (London: Abacus, 1980).

25. The leading examples of this "survivalist school" are Garrett Hardin, *Exploring New Ethics for Survival: The Voyage of the Spaceship Beagle* (Baltimore: Penguin, 1972); William Ophuls, "Leviathan or Oblivion?" in *Toward a Steady State Economy*, ed. Herman E. Daly (San Francisco: Freeman, 1973), 215–30; William Ophuls, *Ecology and the Politics of Scarcity: A Prologue to a Political Theory of the Steady State* (San Francisco: Freeman, 1977); and Robert L. Heilbroner, *An Inquiry into the Human Prospect* (New York: Norton, 1974). It should be noted that Ophul's contribution is particularly wide ranging and eclectic (e.g., he draws on elements of Plato, Aristotle, Benedictine communalism, Hobbes, Rousseau, Burke, Jeffersonian democracy, and Utopian Socialism) and it is possible to find in his work all three of the ecopolitical themes identified in this chapter. Nonetheless, I would characterize his "bottom line" orientation as survivalist. Edward Goldsmith, a key author of *Blueprint* and long time editor of *The Ecologist*, may also be seen as partially belonging to this survivalist school. Although he advocates (unlike the survivalists) the *immediate* transition toward a society made up of decentralized, self-sufficient eco-communities he envisages that such a society would be planned and engineered by the nation state. This idiosyncratic mixture of paternalism, utopianism, and radical conservatism is particularly evident in the final chapter of his recent book *The Great U-Turn: De-Industrializing Society* (Hartland, U.K.: Green Books, 1988) where he emphasizes the importance of traditional, stabilizing institutions (such as the family and religious hierarchies) and rejects public social security institutions in favor of community self-help.

26. Heilbroner, *Human Prospect*, 142–44.

27. Garrett Hardin, "The Tragedy of the Commons," *Science* 162 (1968): 1243–48. Reprinted in K. S. Shrader-Frechette, ed., *Environmental Ethics* (Pacific Grove, Calif.: Boxwood, 1981), 242–52. (All citations refer to this reprint.)

28. That traditional commons were mostly managed on a sustainable basis by local people for mutual benefit (see Susan Jane Buck Cox, "No Tragedy on the Commons," *Environmental Ethics* 7 [1985]: 49–61; and John Reader, "Human Ecology: How Land Shapes Society," *New Scientist*, 8 September 1988, 51–55) does not detract from the force of Hardin's parable in highlighting the "free rider" and Prisoner's Dilemma problems in public choice theory. Moreover, Hardin has replied to John Reader's critique by pointing out that his article was essentially about, and should have been titled, "The Tragedy of the Unmanaged Commons." See Hardin, "Commons Failing," *New Scientist*, 22 October 1988, 76.

29. Hardin, "The Tragedy of the Commons," 250. Hardin's pithy formula ("mutual coercion, mutually agreed upon by the majority of the people affected") is often referred to disparagingly as an apology for a Hobbesian Leviathan yet it has much more in common with the democratic "self-limiting" social contract theory of

Lockean liberalism (with its concern for limited government) than it does with a heavy handed totalitarian state or absolute sovereign.

30. See, for example, Barnet, *The Lean Years*, 297–98.

31. Ibid., 296–97.

32. Heilbroner saw an "ultimate certitude" in environmental destruction, which placed it in an altogether different category from the threat of nuclear war (*Human Prospect*, 47).

33. Ibid., 135.

34. See Ophuls, "Leviathan or Oblivion?" and *Ecology and the Politics of Scarcity*.

35. Ophuls, "Leviathan or Oblivion?" 227.

36. Ophuls, *Ecology and the Politics of Scarcity*, chapter 8.

37. Ibid., 164–65.

38. Ken Walker has argued that Ophuls mistakes Hobbes as identifying material scarcity as the source of social disorder. According to Walker, Hobbes' *Leviathan* was premised on the human desire for eminence, irrespective of whether there was material scarcity. See K. J. Walker, "The Environmental Crisis: A Critique of Neo-Hobbesian Responses," *Polity* 21 (1988): 67–81.

39. Ophuls has framed the central question as follows: "How is the common interest of the collectivity to be achieved when men throughout history have shown themselves to be passionate creatures prey to greed, selfishness and violence?" William Ophuls, "Reversal is the Law of Tao: The Immanent Resurrection of Political Philosophy," in *Environmental Politics*, ed. Stuart S. Nagel (New York: Praeger, 1974), 37. Indeed, Ophuls has heralded the conservative political thinker Edmund Burke to be "the last great spokesman for the premodern point of view" and has endorsed his view that (i) humans are by nature passionate, (ii) there must therefore be checks on will and appetite, and (iii) if these checks are not self-imposed then they must be imposed externally by a sovereign power. See Ophuls, *Ecology and the Politics of Scarcity*, 235.

40. See, for example, Michael E. Kraft, "Analyzing Scarcity: The Politics of Social Change," *Alternatives* Winter, 1978:30–33.

41. Andrew Feenberg, "Beyond the Politics of Survival," *Theory and Society* 7 (1979): 323.

42. Enzensberger, "A Critique of Political Ecology," 15–17.

43. Volkmar Lauber, "Ecology, Politics and Liberal Democracy," *Government and Opposition* 13 (1978): 199–217.

44. See, for example, David W. Orr and Stuart Hill, "Leviathan, the Open Soci-

ety, and the Crisis of Ecology," *The Western Political Quarterly* 31 (1978): 457–69 and Robert Holsworth, "Recycling Hobbes: The Limits to Political Ecology," *The Massachusetts Review* 20 (1979): 9–40. Robert Paehlke has pointed out that at the time Ophuls and Heilbroner wrote their pessimistic theoretical treatises, environmental interest groups were busily *expanding* opportunities for democratic participation in resource management. See Robert Paehlke, "Democracy, Bureaucracy, and Environmentalism," *Environmental Ethics* 10 (1988): 291–308.

45. Susan M. Leeson, "Philosophic Implications of the Ecological Crisis: The Authoritarian Challenge to Liberalism," *Polity* 11 (1979): 305.

46. Ibid., 317.

47. For a general overview see Robert Paehlke, *Environmentalism and the Future of Progressive Politics* (New Haven: Yale University Press, 1989), chapter 7. Although Paehlke notes that environmentalists occupy a wide range of positions on the traditional ideological spectrum, he nonetheless concludes that environmentalism "implies some doubt about the liberal tradition of technocratic management. It suggests that we need to find new means of intervening deeply in the market process—an idea foreign to liberalism and moderate progressivism" (211).

48. The phrase "crisis of culture and character" was introduced into Green discourse by Wendell Berry in *The Unsettling of America* (New York: Avon, 1977).

49. See, for example, Fred Hirsch, *Social Limits to Growth* (Cambridge: Harvard University Press, 1976) and William Leiss, *The Limits to Satisfaction: On Needs and Commodities* (London: Marion Boyars, 1978).

50. Leiss, *The Limits to Satisfaction*, 112.

51. Rodman, "Paradigm Change in Political Science," 72.

52. Bookchin, *Post-Scarcity Anarchism*, 58.

53. Theodore Roszak, *Person/Planet: The Creative Disintegration of Industrial Society* (London: Victor Gollancz, 1979; reprint ed., London: Paladin, 1981), 15.

54. See Rudolf Bahro, "Socialism, Ecology and Utopia: An Interview," *History Workshop* 16 (1983): 94.

55. Christopher Stone, *Should Trees Have Standing?: Toward Legal Rights for Natural Objects* (Los Altos, Calif.: Kaufman, 1974), 48. On this theme, see also Thomas E. Hill, Jr., "Ideals of Human Excellence and Preserving Natural Environments," *Environmental Ethics* 5 (1983): 211–24.

56. Bill Devall and George Sessions, *Deep Ecology: Living as if Nature Mattered* (Layton, Utah: Gibbs M. Smith, 1985), 180.

57. Patrick Hanks, ed. *Collins Dictionary of the English Language* (London: Collins, 1983).

58. See Stephen Rainbow, "Eco-politics in Practice: Green Parties in New

Zealand, Finland and Sweden," Paper presented at the Ecopolitics IV conference, University of Adelaide, South Australia, 21–24 September 1989, 5 (quoting from Values Party, *Blueprint for New Zealand*, 1972, 1). The world's first Green party was the United Tasmania Group, formed in 1972 in the Australian state of Tasmania shortly before the founding of the New Zealand Values Party in the same year. See Pamela Walker, "The United Tasmania Group: An Analysis of the World's First Green Party," in *Environmental Politics in Australia and New Zealand*, ed. Peter Hay, Robyn Eckersley, and Geoff Holloway (Hobart: University of Tasmania, 1989), 161–74.

59. Cornelius Castoriadis, "From Ecology to Autonomy," *Thesis Eleven* no. 3 (1981): 14.

60. For a more general discussion of the relationship of environmentalism to political theory, see P. R. Hay, "Ecological Values and Western Political Traditions: From Anarchism to Fascism," *Politics* (U.K.) 8 (1988): 22–29.

61. Although I have characterized Ophuls's "bottom line" position as survivalist, an emancipatory approach is nonetheless discernible as a sub-theme in his highly eclectic and wide-ranging ecopolitical writings.

62. David Wells, "Radicalism, Conservatism, and Environmentalism," *Politics* 13 (1978): 305.

63. Langdon Winner, *The Whale and the Reactor: A Search for Limits in an Age of High Technology* (Chicago: Chicago University Press, 1986), 57, quoted by Andrew McLaughlin in his review of Winner's book in *Environmental Ethics* 9 (1987): 377.

64. Fred Singleton, "Eastern Europe: Do the Greens Threaten the Reds?," *The World Today* 42 (1986): 160.

65. Leeson, "Philosophic Implications of the Ecological Crisis," 306.

66. Ibid., 305–6.

67. See J. S. Mill, *Principles of Political Economy*, ed. Donald Winch (Harmondsworth, U.K.: Penguin, 1979), chapter 6, 111–17. For a discussion, see John Rodman, "The Liberation of Nature?" *Inquiry* 20 (1977): 115–19.

68. Peter Singer, *Animal Liberation* (New York: The New Review, 1975; reprint ed., New York: Avon, 1975). These ideas are discussed in chapter 2.

69. According to Rodman, J. S. Mill displayed an "ecological sensitivity" in his plea for individuality and diversity and his critique of monoculture, which Rodman interprets as indirectly affirming the intrinsic value of the nonhuman world ("The Liberation of Nature?" 116). Despite these observations, Rodman has provided a lengthy and convincing critique of attempts to extend the liberal notion of "rights" to the nonhuman world. Moreover, he has elsewhere made it clear that liberalism is incompatible with an ecocentric perspective (although he carries forward the liberal principles of diversity and tolerance.) See, for example, John Rodman, "What is Liv-

ing and What is Dead in the Political Philosophy of T. H. Green," *The Western Political Quarterly* 26 (1973): 580.

70. Mark Sagoff has likewise observed that "environmentalism may seem, then, to involve a sort of communitarianism that is inconsistent with principles traditionally associated with a liberal state." See Sagoff's chapter "Can Environmentalists be Liberals?" in *The Economy of the Earth* (Cambridge: Cambridge University Press, 1988), 146–70, especially 147.

71. Rodman, "Paradigm Change in Political Science," 61.

72. See, for example, William R. Catton, Jr. and Riley E. Dunlap, "A New Ecological Paradigm for Post-Exuberant Sociology," *American Behavioral Scientist* 24 (1980): 15–47 and Ophuls, *Ecology and the Politics of Scarcity*, 229.

73. Arne Naess, "The Shallow and the Deep, Long-Range Ecology Movement. A Summary," *Inquiry* 16 (1973): 95–100. For a thorough account of how and why this distinction (which was *not* the first of its kind) has become so influential in ecophilosophical and wider circles, see Warwick Fox, *Toward a Transpersonal Ecology: Developing New Foundations for Environmentalism* (Boston: Shambhala, 1990), especially chapters 2 and 3. See also Timothy O'Riordan, *Environmentalism*, 2d ed. (London: Pion, 1981), 1; Donald Worster, *Nature's Economy: A History of Ecological Ideas* (Cambridge: Cambridge University Press, 1985), xi; Murray Bookchin, *Toward an Ecological Society* (Montreal: Black Rose, 1980), 58–59; and Bookchin, *The Ecology of Freedom* (Palo Alto, California: Cheshire, 1982), 21; William R. Catton, Jr. and Riley E. Dunlap, "A New Ecological Paradigm for Post-Exuberant Sociology"; and Catton and Dunlap, "Environmental Sociology: A New Paradigm," *American Sociologist* 13 (1978): 41–49; Alan R. Drengson, "Shifting Paradigms: From the Technocratic to the Person-Planetary," *Environmental Ethics* 3 (1980): 221–40; and Drengson, *Beyond Environmental Crisis: From Technocrat to Planetary Person* (New York: Peter Lang, 1989). Although the above list indicates something of the pervasiveness of the deep/shallow and comparable distinctions, it is by no means exhaustive. For the most thorough survey and discussion of comparable bipartite distinctions in the ecophilosophical literature to date, see Fox, *Transpersonal Ecology*, 22–40.

74. I prefer *ecocentrism* to *biocentrism* for the reasons given by Warwick Fox in "The Deep Ecology-Ecofeminism Debate and its Parallels," *Environmental Ethics* 11 (1989): 7–8. In particular, the prefix "*eco*" (unlike the prefix "*bio*") encompasses not only individual organisms that are biologically alive but also such things as species, populations, and cultures considered as entities in their own right.

75. Fox argues that it is more accurate to refer to the distinctive philosophical sense of deep ecology as "transpersonal ecology"—rather than deep ecology—since it refers to the realization of a sense of self that includes yet also extends beyond (i.e, is 'trans-') one's egoistic, biographical, or personal sense of self.

76. I use the term *life-forms* throughout this inquiry to include not only individual living organisms but also self-regenerating ecological entities such as populations,

species, ecosystems, and the biosphere. The criterion of self-regeneration, or "autopoiesis," is explained and discussed in chapter 3.

77. As I point out in chapters 6 and 7, ecocentric emancipatory theorists also advocate a range of other redistributive and educational measures; however, reducing births forms a central part of an ecocentric population policy.

78. For example, even bioregionalism—arguably the most politically innovative approach insofar as it argues that human political decision making units should conform to ecological criteria such as watersheds—has not generated new institutional forms. The bioregional case for a patchwork of self-governing communities loosely linked together in a confederation has long been advocated by anarchist theorists.

NOTES TO CHAPTER 2

1. Although the phrase "streams of environmentalism" is now in fairly common usage, I first came across it in Bill Devall's 1979 manuscript "Streams of Environmentalism." A thoroughly revised version of this manuscript appeared as two papers: Devall, "Reformist Environmentalism," *Humboldt Journal of Social Relations* 6 (1979): 129–57; and Devall, "The Deep Ecology Movement," *Natural Resources Journal* 20 (1980): 299–322.

2. John Rodman, "Theory and Practice in the Environmental Movement: Notes Towards an Ecology of Experience," in *The Search for Absolute Values in a Changing World: Proceedings of the Sixth International Conference on the Unity of the Sciences*, vol. 1 (San Francisco: The International Cultural Foundation, 1978), 45–56; Rodman, "Four Forms of Ecological Consciousness Reconsidered," in *Ethics and the Environment*, eds. Donald Scherer and Thomas Attig (Englewood Cliffs, N.J.: Prentice-Hall, 1983), 82–92; Warwick Fox, "Ways of Thinking Environmentally (and Some Brief Comments on their Implications for Acting Educationally)," in *Thinking Environmentally...Acting Educationally: Proceedings of the Fourth National Conference of the Australian Association of Environmental Education*, ed., J. Wilson, G. Di Chiro, and I. Robottom (Melbourne: Victorian Association for Environmental Education, 1986), 21–29; Fox, *Toward a Transpersonal Ecology: Developing New Foundations for Environmentalism* (Boston: Shambhala, 1990), chapter 6. I have also drawn on the overview of environmentalism provided by Devall in "Reformist Environmentalism"; on John Livingston's critique of anthropocentrism in *The Fallacy of Wildlife Conservation* (Toronto: McClelland and Stewart, 1981), especially chapter 2; on J. Baird Callicott's comparison of animal liberation with Leopold's land ethic in "Animal Liberation: A Triangular Affair," *Environmental Ethics* 2 (1980): 311–38; and on John Rodman's critique of animal liberation in "The Liberation of Nature?" *Inquiry* 20 (1977): 83–145.

3. Rodman's typology (as presented in "Four Forms") is (i) resource conservation, (ii) wilderness preservation, (iii) moral extensionism, and (iv) ecological sensibility. Fox, on the other hand, (in *Transpersonal Ecology*, chapters 6 and 7) distinguishes between (i) instrumental value theory, (ii) intrinsic value theory, and (iii) transpersonal

ecology. Fox subdivides instrumental value theory approaches into (i) unrestrained exploitation and expansionism, (ii) resource conservation and development, and (iii) resource preservation. He subdivides intrinsic value theory approaches into (i) ethical sentientism, (ii) biological ethics, (iii) autopoietic ethics (which includes ecosystem ethics and ecosphere—or "Gaian"—ethics), and (iv) cosmic purpose ethics. In contrast to these axiological (i.e., value theory) approaches, transpersonal ecology represents a more phenomenological approach to ecophilosophy. A considerably earlier version of Fox's categories (in "Ways of Thinking Environmentally") has also been used by Alan R. Drengson in "Protecting the Environment, Protecting Ourselves: Reflections on the Philosophical Dimension," in *Environmental Ethics*, vol. 2, ed. R. Bradley and S. Duguid (Vancouver: Simon Fraser University, 1989), 44.

4. I do not discuss the most blatant anthropocentric environmental position, which Fox characterizes as "unrestrained exploitation and expansionism," since no Green activist or emancipatory ecopolitical theorist would support this position.

5. This point is has been developed in P. R. Hay and M. G. Haward, "Comparative Green Politics: Beyond the European Context?" *Political Studies* 36 (1988): 433–48. See also Robyn Eckersley, "Environmental Theory and Practice in the Old and New Worlds: A Comparative Perspective," *Alternatives* (1992), forthcoming.

6. See Alan R. Drengson, "Forests and Forestry Practices: A Philosophical Overview," *Forest Farm Journal* 2 (1990): 3 and Clarence J. Glacken, "The Origins of the Conservation Philosophy," in *Readings in Resource Management and Conservation*, ed. Ian Burton and Robert W. Kates (Chicago: Chicago University Press, 1965), 158.

7. Devall, "Reformist Environmentalism," 140. See also Grant McConnell, "The Environmental Movement: Ambiguities and Meanings," *Natural Resources Journal* 11 (1971): 427–35; Samuel P. Hays, *Conservation and the Gospel of Efficiency* (Cambridge: Harvard University Press, 1959); and Rodman, "Four Forms."

8. Gifford Pinchot, *The Fight for Conservation* (New York: Doubleday Page & Co., 1910), 46. McConnell notes that the principle of development "was in part a reply to those critics who claimed that the goal of conservationists was the mere 'withholding of resources for future generations,' a form of hoarding." However he also notes that, in any event, the principle of development was one to which "the movement under Pinchot's guidance was deeply committed." Grant McConnell, "The Conservation Movement: Past and Present," in *Readings in Resource Management and Conservation*, ed. Ian Burton and Robert W. Kates (Chicago: Chicago University Press, 1965), 191.

9. McConnell, "The Environmental Movement," 430.

10. Devall, "Reform Environmentalism," 140.

11. Rodman, "Four Forms," 83.

12. Neil Evernden, "The Environmentalist's Dilemma," in *The Paradox of Environmentalism*, ed. Neil Evernden (Downsview, Ont.: Faculty of Environmental

Studies, York University, 1984): 10. See also Livingston, *The Fallacy of Wildlife Conservation*, 43–46.

13. Laurence Tribe, "Ways Not to Think About Plastic Trees: New Foundations for Environmental Law," *The Yale Law Journal* 83 (1974): 1315–48. See also Livingston, *The Fallacy of Wildlife Conservation*, 24–34.

14. Friedrich Engels, *The Condition of the Working Class in England*, trans. and ed. W. O. Henderson and W. H. Chaloner, 2d ed. (Oxford: Blackwell, 1971).

15. Robyn Eckersley, "Green Politics and the New Class: Selfishness or Virtue?" *Political Studies* 37 (1989): 205–23. See also Samuel P. Hays, *Beauty, Health and Permanence: Environmental Politics in the United States, 1955–1985* (Cambridge: Cambridge University Press, 1987), 34–35.

16. Livingston, *The Fallacy of Wildlife Conservation*, 34–41.

17. Barry Commoner, *The Closing Circle: Nature, Man and Technology* (New York: Knopf, 1971; reprint ed., New York: Bantam, 1972), 29–44.

18. *Die Grunen, Programme of the German Green Party* (London: Heretic, 1983), 7. Concern for the protection of other species can, however, be found (34–35).

19. Livingston, *The Fallacy of Wildlife Conservation*, 42.

20. Fox, *Transpersonal Ecology*, 186. For a similar discussion, see Evernden, "The Environmentalist's Dilemma."

21. On the difference between conservation and preservation, see John Passmore, *Man's Responsibility for Nature: Ecological Problems and Western Traditions*, 2d ed. (London: Duckworth, 1980), 73 and Rodman, "Four Forms," 84. On the Muir/Pinchot controversy, see Roderick Nash, *Wilderness and the American Mind*, 3d ed. (New Haven: Yale University Press, 1982), especially 135–140.

22. Nash, *Wilderness and the American Mind*, 108.

23. J. G. Mosley, "Toward a History of Conservation in Australia," in *Australia as Human Setting*, ed. Amos Rapaport (Sydney: Angus and Robertson, 1972), 148.

24. The assertion of an ecocentric/biocentric ethic was central to this campaign. See Gary Easthope and Geoff Holloway, "Wilderness as the Sacred: The Franklin River Campaign," in *Environmental Politics in Australia and New Zealand*, ed. by Peter Hay, Robyn Eckersley, and Geoff Holloway (Hobart: Centre for Environmental Studies, University of Tasmania, 1989), 189–201.

25. Stephen Fox, *John Muir and His Legacy* (Boston: Little Brown, 1981), 52–53.

26. Rodman, "Theory and Practice," 51, and "Four Forms," 84–86.

27. J. Baird Callicott, "The Wilderness Idea Revisited," 1991, MS, especially 14–15.

28. For a comprehensive discussion of the case for large ecosystem reservations, see George Sessions, "Ecocentrism and Global Ecosystem Protection." *Earth First!* 21 December 1989, 26–28.

29. Rodman, "The Liberation of Nature?" 112.

30. See the section on "Resource Preservation" in Fox, *Transpersonal Ecology*, chapter 6.

31. For example, Brian Norton has argued that when "we manipulate and control natural processes, we strike at our own freedom, symbolically and actually" (Brian Norton, "Sand Dollar Psychology," *The Washington Post Magazine*, 1 June, 1986, 14) while Mark Sagoff has argued that wilderness is part of America's heritage and forms an important symbol of the nation's character and history (see Mark Sagoff, "On Preserving the Natural Environment," *The Yale Law Journal* 84 [1974]: 205–67).

32. See Thomas E. Hill, Jr., "Ideals of Human Excellence and Preserving Natural Environments," *Environmental Ethics* 5 (1983): 211–24. Hill suggests that indifference to nonsentient nature, while not a moral vice, is nonetheless "likely to reflect either ignorance, a self-importance, or a lack of self-acceptance which we must overcome to have proper humility" (222).

33. I have used the description "animal liberation movement" in view of the popularity of Peter Singer's influential defence of the rights of animals in *Animal Liberation: A New Ethics for Our Treatment of Animals* (New York: The New Review, 1975; reprint ed., New York: Avon, 1975). See also Peter Singer, ed., *In Defence of Animals* (Oxford: Basil Blackwell, 1985).

34. Rodman, "The Liberation of Nature?"; Paul Shepard, "Animal Rights and Human Rites," *North American Review*, Winter 1974, 35–41; Tribe, "Ways Not to Think About Plastic Trees," 1344–45; J. Baird Callicott, "Animal Liberation"; Rodman, "Four Forms," 86–88; John Livingston, "The Dilemma of the Deep Ecologist," in *The Paradox of Environmentalism*, ed. Neil Evernden (Downsview, Ont.: Faculty of Environmental Studies, York University, 1984), 61–72; and Warwick Fox, "Towards a Deeper Ecology?" *Habitat Australia*, August 1985, 26–28.

35. Jeremy Bentham, *An Introduction to the Principles of Morals and Legislation* (1789) chapter 17, quoted by Singer, *Animal Liberation*, 8.

36. In this general overview of the case for animal liberation, I have singled out the particular arguments of Peter Singer in his popular book *Animal Liberation* as this represents the classic, and still the most influential, defence of animal liberation. There are, however, other philosophical justifications for ascribing moral rights to nonhuman animals that do not rest on utilitarianism (although they still acknowledge the importance of sentience). For example, Tom Regan argues for the humane treatment of a more restrictive class of animals ("mentally normal mammals of a year or more") on the basis that they enjoy a mental life of their own and therefore possess "inherent value" and ought to be respected as having moral rights (see Tom Regan, *The Case for Animal Rights* [Berkeley and Los Angeles: University of California Press, 1983] and Regan, "Animal Rights, Human Wrongs," *Environmental Ethics* 2

[1980]: 99–120). In contrast, Stephen Clark places less emphasis on pain and more emphasis on the need to enable animals as well as humans to realize their special potentialities (see S. L. R. Clark, *The Moral Status of Animals* [Oxford: Clarendon Press, 1975]). For a comparison of Clark's and Singer's approach, see John Benson, "Duty and the Beast," *Philosophy* 53 (1978): 529–49.

37. Singer, *Animal Liberation*, 8–9.

38. Ibid., 178–79.

39. Ibid., 7. Richard Ryder, *Speciesism: The Ethics of Vivisection* (Edinburgh: Scottish Society for the Prevention of Vivisection, 1974).

40. Callicott, "Animal Liberation," 313. See also Roderick Nash, "Rounding Out the American Revolution: Ethical Extensionism and the New Environmentalism," in *Deep Ecology*, ed. Michael Tobias (San Diego: Avant, 1985), 170–81 and Roderick Nash, *The Rights of Nature: A History of Environmental Ethics* (Madison: University of Wisconsin Press, 1989).

41. Rodman, "The Liberation of Nature?" 91. See also Callicott, "Animal Liberation," 318.

42. Rodman, "Four Forms," 87. See also Fox, "Towards a Deeper Ecology?" 27.

43. Rodman, "The Liberation of Nature?" 94.

44. Rodman, "Four Forms," 87. See also Callicott, "Animal Liberation."

45. Fox, *Transpersonal Ecology*, 195. J. Baird Callicott has described the issue of predation as the "Achilles' heel" of the case for animal rights put forward by Tom Regan. He argues that while Regan "is not willing to embrace the implications of his theory regarding predators," others, such as Steve Sapontzis, have been more forthright. In particular, Sapontzis has argued that it would be a morally better world if there were no carnivores at all. See J. Baird Callicott, review of *The Case for Animal Rights*, by Tom Regan, *Environmental Ethics* 7 (1985): 371 and Steve S. Sapontzis, "Predation," *Ethics and Animals* 5 (1984): 27–36.

46. Singer, *Animal Liberation*, 238–39. In particular, Singer rejects the idea of policing nonhuman carnivores in the wild on the grounds that (i) any attempt to change ecological systems on a large scale would do more harm than good, and (ii) we ought not to claim dominion over other species: "Having given up the role of tyrant, we should not try to play Big Brother" (ibid). It should be noted, however, that these objections are ad hoc in that they do not flow from Singer's sentience criterion.

47. For a general defence of ecocentric holism, see Peter S. Wenz, *Environmental Justice* (Albany: State University of New York Press, 1988), 292–309.

NOTES TO CHAPTER 3

1. For a clear exposition of this model, see Charles Birch and John B. Cobb, Jr., *The Liberation of Life: From the Cell to the Community* (Cambridge: Cambridge Uni-

versity Press, 1981). See also J. Baird Callicott, "The Metaphysical Implications of Ecology," *Environmental Ethics* 8 (1986): 301–16 and Andrew McLaughlin, "Images and Ethics of Nature," *Environmental Ethics* 7 (1985): 293–19.

2. Birch and Cobb, *The Liberation of Life*, 95.

3. See Richard Routley and Val Routley, "Human Chauvinsim and Environmental Ethics," in *Environmental Philosophy*, eds. Don Mannison, Michael McRobbie, and Richard Routley, Monograph Series, No. 2, Department of Philosophy, Research School of the Social Sciences, Australian National University, Canberra, Australia, 96–189. See also Warwick Fox, *Toward a Transpersonal Ecology: Developing New Foundations for Environmentalism* (Boston: Shambhala, 1990), 15–17.

4. Fox, *Transpersonal Ecology*, 15.

5. R. Routley and V. Routley, "Against the Inevitability of Human Chauvinism," in *Ethics and Problems of the 21st Century*, eds. K. E. Goodpaster and K. M. Sayre (Notre Dame: University of Notre Dame Press, 1979), 36–59. For an excellent discussion of supposed human-animal discontinuities, see Barbara Noske, *Humans and Other Animals* (London: Pluto, 1989), chapter 6.

6. John Rodman, "Paradigm Change in Political Science," *American Behavioral Scientist* 24 (1980): 54.

7. Ted Benton, "Humanism = Speciesism: Marx on Humans and Animals," *Radical Philosophy* Autumn 1988: 11. Benton is summarizing here an argument of Mary Midgley's from *Animals and Why They Matter* (Harmondsworth, U.K.: Penguin, 1983), chapter 2.

8. George Sessions, "Anthropocentrism and the Environmental Crisis," *Humboldt Journal of Social Relations* 2 (1974): 73.

9. See Fritjof Capra, *The Turning Point: Science, Society, and the Rising Culture* (London: Fontana, 1983; reprint ed., 1985) and Rupert Sheldrake, *The Rebirth of Nature: The Greening of Science and God* (London: Century, 1990). As George Sessions has pointed out ("Ecocentrism and the Greens: Deep Ecology and the Environmental Task," *The Trumpeter* 5 [1988]: 67), the idea of a hierarchical chain of being can be traced back to Aristotle, who "rejected the Presocratic ideas of an infinite universe, cosmological and biological evolution, and heliocentrism, and proposed instead an Earth-centered, finite universe, wherein humans were differentiated from, and seen as superior to, the rest of the animals by virtue of their rationality. Also found in Aristotle is the hierarchical concept of the 'great chain of being' which holds that Nature made plants for the use of animals, and animals were made for the sake of humans (*Politics I*, 88)." Sessions notes that the Presocratics, on the other hand, had been much more interested in cosmological inquiry and nature in general.

10. For three sustained critiques of anthropocentrism, see Routley and Routley, "Against the Inevitability of Human Chauvinism"; David Ehrenfeld, *The Arrogance of Humanism* (New York: Oxford University Press, 1981); and Fox, *Transpersonal Ecology*, 13–22. Fox concludes that anthropocentrism is not only self-serving but also

"empirically bankrupt and theoretically disastrous, practically disastrous, logically inconsistent, morally objectionable, and incongruent with a genuinely open approach to experience" (18–19).

11. Rodman, "Paradigm Change in Political Science," 67; Capra, *The Turning Point*; Sessions, "Anthropocentrism and the Environmental Crisis"; and Sessions, "Ecocentrism and the Greens."

12. Warwick Fox, "Deep Ecology: A New Philosophy of Our Time?" *The Ecologist* 14 (1984): 194–200. See also J. Baird Callicott, "Intrinsic Value, Quantum Theory, and Environmental Ethics," *Environmental Ethics* 7 (1985): 257–75 and Capra, *The Turning Point*.

13. For Capra's somewhat deterministic conclusions, see *The Turning Point*, 464–66.

14. Michael Zimmerman, "Quantum Theory, Intrinsic Value, and Panentheism," *Environmental Ethics* 10 (1988): 5.

15. Michael Zimmerman, "Marx and Heidegger on the Technological Domination of Nature," *Philosophy Today* 23 (1979): 103.

16. The complexity and unpredictability of many physical and social phenomena are underscored by the new body of scientific inquiry known as chaos theory, which shows that dynamic biological and social systems that behave deterministically (i.e., according to laws that can be described mathematically) are nonetheless inherently unpredictable beyond a certain point. This is due to the fact that these systems exhibit nonlinear dynamical properties (which means that they are extraordinarily sensitive to initial conditions) together with the fact that it is impossible in principle to specify the initial conditions of any system precisely. That is, some degree of approximation is always involved. For a general introduction, see James Gleick, *Chaos: Making a New Science* (New York: Viking, 1987).

17. Warwick Fox, "The Deep Ecology-Ecofeminism Debate and its Parallels," *Environmental Ethics* 11 (1989): 6. The term *unfold* is used here and throughout this inquiry to mean "develop" or "grow" and is not intended imply any predestination.

18. Trevor Blake, "Ecological Contradiction: The Grounding of Political Ecology," *Ecopolitics II Proceedings* (Hobart: Centre for Environmental Studies, University of Tasmania, 1987), 79.

19. Birch and Cobb, *The Liberation of Life*, 95.

20. For more on the concept of relative autonomy, see Warwick Fox, *Approaching Deep Ecology: A Response to Richard Sylvan's Critique of Deep Ecology*, Environmental Studies Occasional Paper no. 20 (Hobart: Centre for Environmental Studies, University of Tasmania, 1986), section 3.

21. J. Baird Callicott, "What's Wrong with the Case for Moral Pluralism," Paper presented to the Pacific Division Meeting of the American Philosophy Association, Berkeley, 23 March 1989, 32–33.

22. Evelyn Fox Keller, *Reflections on Gender and Science* (New Haven: Yale University Press, 1985), 99.

23. Zimmerman, "Quantum Theory, Intrinsic Value, and Panentheism," 17.

24. George Sessions, "The Deep Ecology Movement: A Review," *Environmental Review* 11 (1987): 105.

25. Fox, *Transpersonal Ecology*, 21.

26. Routley and Routley, "Against the Inevitaility of Human Chauvinsim," 36 and Fox, *Transpersonal Ecology*, 21.

27. I am grateful to Alan Drengson for drawing my attention to these practices.

28. Noske, *Humans and Other Animals*, 160.

29. Fox, *Transpersonal Ecology*, 21.

30. Fox calls this misinterpretation "the fallacy of misplaced misanthropy" (*Transpersonal Ecology*, 19).

31. See, for example, Sessions, "Anthropocentrism and the Environmental Crisis."

32. H. J. Blackham, *Humanism* (New York: International Publishing Service, 1976), 102.

33. For a discussion of how respect for humans rights can nest within an ecocentric framework, see Peter S. Wenz, *Environmental Justice* (Albany: State University of New York Press, 1988).

34. For an example of a nonanthropocentric intrinsic value approach that seeks to maximize richness of experience while taking into account populations and ecosystems, see Birch and Cobb, *The Liberation of Life*, especially 173–74.

35. See Naess, "The Shallow and the Deep, Long-Range Ecology Movement. A Summary," *Inquiry* 16 (1973): 95.

36. See Bill Devall, *Simple in Means, Rich in Ends: Practicing Deep Ecology* (Layton, Utah: Gibbs M. Smith, 1988).

37. Christopher Stone, *Should Trees Have Standing?: Toward Legal Rights for Natural Objects* (Los Altos, Calif.: Kaufmann, 1974).

38. Ibid., 34.

39. Ibid., 40.

40. Paul Shepard, "Animal Rights and Human Rites," *North American Review*, Winter 1974, 35.

41. John Livingston, *The Fallacy of Wildlife Conservation* (Toronto: McClelland and Stewart, 1981), 62–63.

42. Rodman, "The Liberation of Nature?" *Inquiry* 20 (1970): 101.

43. John Rawls, *A Theory of Justice* (London: Oxford University Press, 1976), 512.

44. New Zealand is in the forefront of comprehensive environmental legislation of this kind. For example, the preamble to the New Zealand Environment Act 1986 states that the purpose of the Act is, inter alia, to "ensure that, in the management of natural and physical resources, full and balanced account is taken of (i) the intrinsic value of ecosystems; and (ii) all values which are placed by individuals and groups on the quality of the environment; and (iii) the principles of the Treaty of Waitangi [i.e., an agreement between White settlers and Maories]; and (iv) the sustainability of natural and physical resources; and (v) the needs of future generations." A further example is the New Zealand Conservation Act 1987, which defines conservation to mean "the preservation and protection of natural and historic resources *for the purpose of maintaining their intrinsic values*, providing for their appreciation and recreational enjoyment by the public, and safeguarding the options of future generations" (section 2[1]—my emphasis). "Natural resources" are defined in the Act to include not only plants and animals, but landscapes and landforms, geological features and "systems of interacting living organisms, and their environment."

45. Charles Elton, the founder of modern animal ecology, has bluntly stated that "the balance of nature does not exist and perhaps never has existed" (Charles Elton, *Animal Ecology and Evolution* [Oxford: Oxford University Press, 1930], 17, quoted in Birch and Cobb, *The Liberation of Life*, 36–37). Birch and Cobb suggest that it is more precise to speak of certain kinds of activity as being "unsustainable" rather than as upsetting the "balance of nature," since the latter suggests that nature is static, that is, that the distribution and abundance of plants and animals in a community does not change. See also Frank N. Egerton, "Changing Concepts of the Balance of Nature," *Quarterly Review of Biology* 48 (1978): 322–50.

46. Livingston, *The Fallacy of Wildlife Conservation*, 75.

47. Elsewhere I have been critical of this tendency in the work of Murray Bookchin (see Robyn Eckersley, "Divining Evolution: The Ecological Ethics of Murray Bookchin," *Environmental Ethics* 11 [1989]: 107). Indeed, ecoanarchism in general is prone to this kind of reasoning, as I show in chapter 7.

48. Although I do not discuss Taoism, Buddhism, and animistic cosmologies, it should be pointed out that autopoietic intrinsic value theory, transpersonal ecology, and ecofeminism are obviously broadly sympathetic with these non-Western approaches. On Taoist and Buddhist approaches, see Ip Po-Keung, "Taoism and the Foundations of Environmental Ethics," *Environmental Ethics* 5 (1983): 335–43, and McLaughlin, "Images and Ethics of Nature," 293–319; and on animistic cosmologies, see J. Donald Hughes, *American Indian Ecology* (El Paso, Tex.: Texas Western Press, 1983) and J. Baird Callicott, "Traditional American Indian and Western European Attitudes Toward Nature: An Overview," *Environmental Ethics* 4 (1982): 293–318.

49. The concept of "autopoiesis" derives from the biological work of Francisco Varela, Humberto Maturana, and Ricardo Uribe. See Francisco J. Varela, Humberto R.

Maturana, and Ricardo Uribe, "Autopoiesis: The Organization of Living Systems, Its Characterization and a Model," *Biosystems* 5 (1974): 187–96, and Humberto R. Maturana and Francisco J. Varela, *The Tree of Knowledge: The Biological Roots of Human Understanding* (Boston: Shambhala, 1988). Fox is responsible for introducing this idea to the environmental philosophy literature (see *Transpersonal Ecology*, 165–76).

50. Fox, *Transpersonal Ecology*, 171.

51. Ibid., 171–72.

52. Ibid., 172. On the meaning of intrinsic value, see also William Godfrey-Smith, "The Value of Wilderness," *Environmental Ethics* 1 (1979): 309.

53. Aldo Leopold, *A Sand County Almanac* (Oxford: Oxford University Press, 1949), 224–25. See also J. Baird Callicott, "The Conceptual Foundations of the Land Ethic," in *Companion to A Sand County Almanac*, ed. J. Baird Callicott (Madison: University of Wisconsin Press, 1987), 186–217 and James D. Heffernan, "The Land Ethic: A Critical Appraisal," *Environmental Ethics* 4 (1982): 235–47.

54. Defenders of Leopold's land ethic have sought to get around this problem by presenting the ethic as a much needed *addition*, rather than alternative, to atomistic approaches to intrinsic value theory such as animal liberation or life-based ethics. See, for example, Callicott, "The Conceptual Foundations of the Land Ethic," 207.

55. Fox, *Transpersonal Ecology*, 169.

56. Bill Devall and George Sessions, *Deep Ecology: Living as if Nature Mattered* (Layton, Utah: Gibbs M. Smith, 1985) and Fox, *Transpersonal Ecology*, especially chapters 7 and 8. In the following discussion I draw on many of the categories and arguments presented by Fox in *Transpersonal Ecology* on the differences between intrinsic value theory approaches and psychological-cosmological approaches (these approaches are explained in the text).

57. For a helpful discussion on the distinction between acting according to moral principle and acting "according to the heart," see Alan R. Drengson, "Compassion and Transcendence of Duty and Inclination," *Philosophy Today* Spring (1981): 34–45.

58. See Elizabeth M. Pybus and Alexander Broadie, "Kant and the Maltreatment of Animals," *Philosophy* 53 (1978): 560–61.

59. Warwick Fox, "The Meanings of 'Deep Ecology,'" *Island Magazine*, Autumn 1989, 34 (this article has also appeared in *The Trumpeter* 7 [1990]: 48–50).

60. Arne Naess, "Self-realization: An Ecological Approach to Being in the World," *The Trumpeter* 4 (1987): 39.

61. Fox, *Transpersonal Ecology*, chapter 7.

62. Transpersonal ecology should *not* be confused with a "New Age" perspective. Indeed, deep/transpersonal ecologists have been quite critical of New Age ideas, particularly those of the Christian theologian Pierre Teilhard de Chardin. See George Sessions, review of *The Soul of the World: An Account of the Inwardness of Things*,

by Conrad Bonifazi, *Environmental Ethics* 3 (1981): 275–81; Sessions, review of *Eco-Philosophy: Designing New Tactics for Living*, by Henryk Skolimowski, *Environmental Ethics* 6 (1984): 167–74; and Devall and Sessions, *Deep Ecology*, 5–6 and 138–44. For the historical roots of the transpersonal ecology approach one needs to look in the direction of people as diverse (in some senses) as Spinoza and Gandhi (see Fox, *Transpersonal Ecology*, chapter 4).

63. Alan R. Drengson, "Developing Concepts of Environmental Relationships," *Philosophical Inquiry* 8 (1986): 50–65.

64. Lester Milbrath, *Environmentalists: Vanguard for a New Society* (Albany: State University of New York Press, 1984), 28.

65. Fox, *Transpersonal Ecology*, 247.

66. See Warwick Fox, "New Philosophical Directions in Environmental Decision-making," in *Theoretical Issues in Environmentalism: Essays from Australia*, eds. P. R. Hay, R. Eckersley, and G. Holloway (Hobart: University of Tasmania, 1991), forthcoming.

67. See, for example, the New Zealand Environment Act 1986 and Conservation Act 1987, discussed above.

68. For general introductions to ecofeminism, see Rosemary Radford Ruether, *New Woman New Earth: Sexist Ideologies and Human Liberation* (New York: Seabury, 1975); Susan Griffin, *Woman and Nature: The Roaring Inside Her* (New York: Harper and Row, 1978); Elizabeth Dodson Gray, *Green Paradise Lost* (Wellesley, Mass.: Roundtable, 1981); Brian Easlea, *Science and Sexual Oppression: Patriarchy's Confrontation with Woman and Nature* (London: Weidenfeld and Nicolson, 1981); Carolyn Merchant, *The Death of Nature: Women, Ecology and the Scientific Revolution* (London: Wildwood House, 1982); Isaac D. Balbus, *Marxism and Domination: A Neo-Hegelian, Feminist, Psychoanalytical Theory of Sexual, Political and Technological Liberation* (Princeton, N.J.: Princeton University Press, 1982); Leonie Caldecott and Stephanie Leland, eds., *Reclaim the Earth: Women Speak Out for Life on Earth* (London: Women's Press, 1983); Joan Rothschild, ed., *Machina Ex Dea: Feminist Perspectives on Technology* (New York: Pergamon, 1983); Val Plumwood, "Ecofeminism: An Overview and Discussion of Positions and Arguments," *Australasian Journal of Philosophy* 64 (1986): 120–38; Patsy Hallen, "Making Peace with Nature: Why Ecology Needs Feminism," *The Trumpeter* 4 (1987): 3–14; Val Plumwood, "Women, Humanity and Nature," *Radical Philosophy* (Spring 1988): 16–24; Judith Plant, ed., *Healing the Wounds: The Promise of Ecofeminism* (Philadelphia: New Society Publishers, 1989); Irene Diamond and Gloria Feman Orenstein, eds., *Reweaving the World: The Emergence of Ecofeminism* (San Francisco: Sierra Club Books, 1990); Karen J. Warren, "Feminism and Ecology: Making Connections," *Environmental Ethics* 9 (1987): 3–20; and also by Warren, "The Power and the Promise of Ecological Feminism," *Environmental Ethics* 12 (1990): 125–46.

69. Irene Diamond and Gloria Feman Orenstein, "Introduction," *Reweaving the World*, ix. Similarly, Ynestra King has argued that the domination of women is "the

prototype of other forms of domination" of which the domination of nature is but one example. Ynestra King, "Toward an Ecological Feminism and a Feminist Ecology," in *Machina Ex Dea,* ed., Rothschild, 119.

70. Simone de Beauvoir, *The Second Sex,* trans. and ed. H. M. Parshley (New York: Knopf, 1978; reprint ed., Harmondsworth, U.K.: Penguin, 1982). Although de Beauvoir was a feminist existentialist and not an *eco*feminist (indeed, she rejected the association of women with nature on the grounds that it inhibited women's own process of becoming free, independent existents), her observations on the relationship between woman, man, and nature in *The Second Sex* have been widely drawn upon by contemporary ecofeminist theorists.

71. See, for example, Charlene Spretnak, *The Spiritual Dimension of Green Politics* (Santa Fe, N. Mex.: Bear and Company, n.d.); reprinted as Appendix C in Charlene Spretnak and Fritjof Capra, *Green Politics: The Global Promise* (London: Paladin, 1986), 230–58; Starhawk, *Dreaming the Dark: Magic, Sex, and Politics* (Boston: Beacon, 1982); and Judith Plant, ed., *Healing the Wounds,* part 3 ("She is Alive in You: Ecofeminist Spirituality"), 115–88.

72. Gary Snyder, "Anarchism, Buddhism, and Political Economy," lecture delivered at the Fort Mason Center, San Francisco, 27 February 1984 (quoted by Charlene Spretnak, *The Spiritual Dimensions of Green Politics,* in Spretnak and Capra, *Green Politics,* 238.)

73. Dodson Gray, *Green Paradise Lost,* 148.

74. Ibid., 84 and 85.

75. Ariel Kay Salleh, "Deeper than Deep Ecology: The Ecofeminist Connection," *Environmental Ethics* 6 (1984): 339–45; Jim Cheney, "Eco-feminism and Deep Ecology," *Environmental Ethics* 9 (1987): 115–45; and Marti Kheel, "Ecofeminism and Deep Ecology: Reflections on Identity and Difference," in *Reweaving the World,* ed. Irene Diamond and Gloria Feman Orenstein, 128–37.

76. Fox, *Transpersonal Ecology,* 258.

77. Ibid., 262–63. See also Fox, "The Deep Ecology-Ecofeminism Debate," 11–12.

78. See Fox, "The Deep Ecology-Ecofeminism Debate," 12–13. See also Dolores LaChapelle, *Sacred Land, Sacred Sex: Rapture of the Deep* (Silverton, Col.: Finn Hill Arts, 1988).

79. Salleh, "Deeper than Deep Ecology," 342–43; King, "Toward an Ecological Feminism," 123; and Cheney, "Eco-feminism and Deep Ecology."

80. Elizabeth Dodson Gray has gone so far as to claim "that there is a definite limit to the perception of men. It is a limit imposed upon their consciousness by the lack of certain bodily experiences which are present in the life of woman.... the male's is simply a much diminished experience of body, of natural processes, and of future generations" (Gray, *Green Paradise Lost,* 113–114). See also Salleh, "Deeper than Deep Ecology," 340.

81. See, for example, Spretnak, "The Spiritual Dimension of Green Politics"; LaChapelle, *Earth Wisdom* (Los Angeles: Guild of Tutors Press, 1978); LaChapelle, *Sacred Land, Sacred Sex*; and John Seed, Joanna Macy, Pat Fleming, and Arne Naess, *Thinking like a Mountain: Towards a Council of All Beings* (Santa Cruz, Calif.: New Society Publishers, 1988).

82. Joan L. Griscom, "On Healing the Nature/History Split in Feminist Thought," *Heresies* 13 (1981): 8.

83. Object relations theory is a branch of psychoanalytic theory concerned with the development of the self in relation to others. For two feminist approaches, see Dorothy Dinnerstein, *The Mermaid and the Minotaur: Sexual Arrangements and Human Malaise* (New York: Harper and Row, 1977) and Nancy Chodorow, *The Reproduction of Mothering* (Berkeley and Los Angeles: University of California Press, 1978). Ecofeminists who draw on object relations theory to defend a "feminine sense of self" include Isaac D. Balbus, "A Neo-Hegelian, Feminist, Psychoanalytic Perspective on Ecology," *Telos* 52 (1982): 140–55; Balbus, *Marxism and Domination*; and Marti Kheel, "Ecofeminism and Deep Ecology," 130–31.

84. See, for example, Keller, *Reflections on Gender and Science*, 89.

85. Carol Gilligan, *In a Different Voice* (Cambridge: Harvard University Press, 1982.

86. See, for example, Ariel Salleh, "Deeper than Deep Ecology." For replies to this charge, see Michael E. Zimmerman, "Feminism, Deep Ecology, and Environmental Ethics," *Environmental Ethics* 9 (1987): 21–44; Warwick Fox, "The Deep Ecology-Ecofeminism Debate"; and Alan E. Wittbecker, "Deep Anthropology: Ecology and Human Order," *Environmental Ethics* 8 (1986): 261–70.

87. See Joan Babberger, "The Myth of Matriarchy: Why Men Rule in Primitive Societies," in *Woman, Culture, and Society*, ed. Michelle Zimbalist Rosaldo and Louise Lamphere (Stanford: Stanford University Press, 1974), 263–80 and Marilyn French, *Beyond Power: Women, Men and Morality* (London: Abacus, 1986), 96–100.

88. See Friedrich Engels, *The Origin of the Family, Private Property and the State*, trans. Alick West and Dona Torr (London: Lawrence and Wishart, 1940; reprint ed., 1946).

89. Fox, "The Deep Ecology-Ecofeminism Debate," 21–25.

90. Plumwood, "Women, Humanity and Nature," 18.

91. See, for example, Sessions, "Anthropocentrism and the Environmental Crisis."

92. Judith Plant, "Introduction," *Healing the Wounds*, 3.

93. Don E. Marietta, "Environmentalism, Feminism, and the Future of American Society," *The Humanist* 44 (1984): 18. Other theorists sympathetic with this kind of approach include Warwick Fox, Joan Griscom, Patsy Hallen, Evelyn Fox Keller, Val Plumwood, Karen Warren, Alan Wittbecker, and Michael Zimmerman.

94. Plumwood, "Women, Humanity and Nature," 22.

95. Ibid., 23.

96. Warren, "The Power and the Promise of Ecological Feminism."

NOTES TO CHAPTER 4

1. See, for example, Hans Magnus Enzensberger, "A Critique of Political Ecology," *New Left Review* 84 (1974): 3–31 and Melanie Beresford, "Doomsayers and Eco-nuts: A Critique of the Ecology Movement," *Politics* 12 (1971): 98–106.

2. It will suffice to point out that many of these Marxist critiques proceed on the basis of (i) an ill-informed understanding of the social composition of the environmental movement; (ii) a crude class model of society (i.e., middle-class capitalists versus the working class) that ignores many significant cleavages *within* the middle class between say, the New Class and the "business class"; and (iii) a characterization of environmentalism that either ignores or fails to grasp the radical implications of much environmental protest. See Robyn Eckersley, "The Environment Movement as Middle-Class Elitism: A Critical Analysis," *Regional Journal of Social Issues* 18 (1986): 24–36 and Eckersley, "Green Politics and the New Class: Selfishness or Virtue?" *Political Studies* 37 (1989): 205–23.

3. Two prominent examples in West Germany are Thomas Ebermann and Rainer Trampert. For a discussion of the relationship between Marxism and *Die Grunen*, see John Ely, "Marxism and Green Politics in West Germany," *Thesis Eleven* no. 13 (1986): 22–38 and Werner Hülsberg, *The German Greens: A Social and Political Profile* (London: Verso, 1988).

4. Major ecopolitical critiques of Marxism include Michael Zimmerman, "Marx and Heidegger on the Technological Domination of Nature," *Philosophy Today* 23 (1979): 99–112; Val Routley, "On Karl Marx as an Environmental Hero," *Environmental Ethics* 3 (1981): 237–44; Isaac D. Balbus, *Marxism and Domination: A Neo-Hegelian, Feminist, Psychoanalytical Theory of Sexual, Political, and Technological Liberation* (Princeton, N.J.: Princeton University Press, 1982); Rudolf Bahro, *Socialism and Survival* (London: Heretic Books, 1982); Hwa Yol Jung, "Marxism, Ecology, and Technology," *Environmental Ethics* 5 (1983): 169–71; M. R. Redclift, "Marxism and the Environment: A View from the Periphery," in *Political Action and Social Identity: Class, Locality and Ideology*, eds. Gareth Rees, Janet Bujra, Paul Littlewood, Howard Newby, and Teresa L. Rees (London: Macmillan, 1985), 191–211; Ted Benton, "Humanism = Speciesism: Marx on Humans and Animals," *Radical Philosophy* Autumn 1988: 4–18; and John Clark, "Marx's Inorganic Body," *Environmental Ethics* 11 (1989): 243–58.

5. For example, Howard Parsons, *Marx and Engels on Ecology* (Westport, Conn.: Greenwood, 1978); Donald Lee, "On the Marxian View of the Relationship between Man and Nature," *Environmental Ethics* 2 (1980): 3–16; and Michael Clow, "Alienation from Nature: Marx and Environmental Politics," *Alternatives* 10 (1982): 36–40.

6. For example, Parsons, *Marx and Engels on Ecology*; Lee, "On the Marxian View"; Charles Tolman, "Karl Marx, Alienation, and the Mastery of Nature," *Environmental Ethics* 3 (1981): 63–74; Raymond Williams, *Socialism and Ecology* (London: Socialist Environment and Resources Association, n.d.); Janna Thompson, "The Death of a Contradiction: Marxism, the Environment and Social Change," *Intervention* 17 (1983): 7–26; Adrienne Farago, "Environmentalism and the Left," *Urban Policy and Research* 3 (1985): 11–15; Neil Smith, *Uneven Development: Nature, Capital and the Production of Space* (Oxford: Blackwell, 1984); Michael Redclift, *Sustainable Development: Exploring the Contradictions* (London: Methuen, 1987), 45–51 and 173–80; James O'Connor, "Capitalism, Nature, Socialism: A Theoretical Introduction," *Capitalism, Nature, Socialism* 1 (1988): 11–38; and James O'Connor, "Introduction to Issue Number Two: Socialism and Ecology," *Capitalism, Nature, Socialism* 2 (1989): 5–11.

7. André Gorz, *Ecology as Politics*, trans. Patsy Vigderman and Jonathan Cloud (London: Pluto, 1980); Gorz, *Farewell to the Working Class: An Essay in Post-Industrial Socialism* (London: Pluto, 1982); Bahro, *Socialism and Survival;* and Bahro, *Building the Green Movement* (London: Heretic/GMP, 1986). Indeed, Rudolf Bahro and Murray Bookchin (also a former Marxist sympathizer) have become two of Marxism's staunchest ecological critics. See, for example, Bahro, *Socialism and Survival* and Murray Bookchin, "Marxism as Bourgeois Sociology," and "On Neo-Marxism, Bureaucracy and the Body Politic," in *Toward an Ecological Society* (Montreal: Black Rose, 1980), 193–210 and 213–48.

8. See Alfred Schmidt, *The Concept of Nature in Marx* (London: New Left Books, 1971), 17.

9. Ibid., 154.

10. Ibid., 15 and 81.

11. Ibid., 79.

12. Karl Marx, *The Economic and Philosophical Manuscripts of 1844*, trans. Martin Milligan, ed. Dirk J. Struik (New York: International Publishers, 1964), 112. This work was not published in full in Europe until the 1930s and did not become generally available in the United States until the 1960s.

13. Feuerbach had argued in his famous "inversion" of Hegel's idealism that the "subject" of history was neither the Absolute Idea nor the individual self but rather nature rendered self-conscious in humanity taken as a whole. Feuerbach was both an atheist and a materialist who regarded God as an alien and fictitious being to whom humans had attributed their essential powers, thereby impoverishing themselves. Religion and Hegel's idealist metaphysics were therefore both regarded as stages to be overcome in humanity's emergence out of nature since Feuerbach regarded humanity, not God, as the appropriate object of worship. According to Feuerbach, we realize our species being or human essence as our consciousness of ourselves expands, culminating in the overcoming of our alienation from ourselves and the realization of the unity of subject and object—humanity and nature. Marx, who consid-

ered Feuerbach's conception of humanity's relationship to nature to be too passive, considerably reworked Feuerbach's notion of self-estrangement on the basis of a different conception of human essence that was based on the dynamics of the labor process. Hegel's *Geist* or spirit was thus replaced by the concrete activity of *homo faber*. For a general discussion of Feurbach's ideas and their influence on the young Marx, see Dirk J. Struik's "Introduction" to *The Economic and Philosophical Manuscripts of 1844*, 15 and following.

14. Marx, *The Economic and Philosophical Manuscripts of 1844*, 113.

15. Ibid, 114.

16. See William James Booth, "Gone Fishing: Making Sense of Marx's Concept of Communism," *Political Theory* 17 (1989): 205–22.

17. Schmidt, *The Concept of Nature in Marx*, 76.

18. According to Schmidt, this is the fundamental distinction between a materialist and idealist dialectic: "In the Marxist dialectic, as in the Hegelian, what is nonidentical with the Subject is overcome stage by stage. Greater and greater areas of nature come under human control. In Marx, however, and this distinguishes him from Hegel's ultimate idealism, the material of nature is never totally incorporated in the modes of its theoretical-practical appropriation" (ibid., 136).

19. Ibid., 129. See Karl Marx and Frederick Engels, *The German Ideology*, trans. by Clemens Dutt, W. Lough, and C. P. Magill, in Karl Marx and Frederick Engels, *Collected Works*, vol. 5 (London: Lawrence & Wishart, 1976). One important exception to this generalization is *The Grundrisse*.

20. See, for example, Karl Marx, *Grundrisse: Foundations of the Critique of Political Economy*, trans. by Martin Nicholas (London: Allen Lane, New Left Review, 1973; reprint ed., New York: Vintage, 1973), 409–10.

21. Karl Marx, *Capital: A Critique of Political Economy*, ed. Friedrich Engels, trans. Samuel Moore and Edward Aveling, vol. 1: *Capitalist Production* (London: Lawrence & Wishart, 1970), 513.

22. Isaac D. Balbus, *Marxism and Domination*, 272.

23. Marx, *Capital*, vol. 1., 179.

24. Karl Marx, *Capital: A Critique of Political Economy*, ed. Friedrich Engels, trans. Samuel Moore and Edward Aveling, vol. 3: *The Process of Capitalist Production as a Whole* (London: Lawrence & Wishart, 1970), 820.

25. Friedrich Engels, "Socialism: Utopian and Scientific" in *The Marx-Engels Reader*, ed. Robert C. Tucker (New York: Norton, 1972), 637–38.

26. Friedrich Engels, *Dialectics of Nature*, trans. Clemens Dutt, in Karl Marx and Friedrich Engels, *Collected Works*, vol. 25 (London: Lawrence & Wishart, 1987), 460.

27. Ibid., 460–61.

28. Ibid., 461.

29 . Marx, *Capital,* vol. 1, 264–65 and 506 and Marx, *Capital,* vol. 3, 812–13.

30. Marx, *Capital,* vol. 1, 505–7.

31. See Tolman, "Karl Marx, Alienation, and the Mastery of Nature"; Parsons, *Marx and Engels on Ecology*; and Beresford, "Doomsayers and Eco-nuts."

32. Tolman, "Karl Marx, Alienation, and the Mastery of Nature," 73.

33. Parsons, *Marx and Engels on Ecology,* 44.

34. Ibid., 45.

35. Ibid.

36. Ibid, 47.

37. Ibid, 45.

38. Ibid., 16.

39. Ibid., 18–19.

40. Clark, "Marx's Inorganic Body," 245.

41. For a juxtaposition of the respective concerns and philosophies of Marx and Muir (and their successors), see Frances Moore Lappé and J. Baird Callicott, "Marx Meets Muir: Toward a Synthesis of the Progressive Political and Ecological Visions," *Tikkun* 2/3 (1987): 16–21 and Robyn Eckersley, "The Road to Ecotopia?: Socialism Versus Environmentalism," *Island Magazine,* Spring 1987, 18–25; reprinted in *The Ecologist* 18 (1988): 142–47; *The Trumpeter* 5 (1988): 60–64; and in *First Rights: A Decade of Island Magazine,* eds. Andrew Sant and Michael Denholm (Elwood, Victoria: Greenhouse, 1989), 50–60.

42. Friedrich Engels, *The Condition of the Working Class in England,* trans. and ed. W. O. Henderson and W. H. Chaloner, 2d ed. (Oxford: Blackwell, 1971).

43. Bookchin, "Marxism as Bourgeois Sociology," 195.

44. Clark, "Marx's Inorganic Body," 258.

45. Parsons, *Marx and Engels on Ecology,* 14.

46. See, for example, Bookchin, *Toward an Ecological Society*; Bahro, *Socialism and Survival*; John Clark, *The Anarchist Moment: Reflections on Culture, Nature and Power* (Montreal: Black Rose, 1984); and Carl Boggs, *Social Movements and Political Power: Emerging Forms of Radicalism* (Philadelphia: Temple University Press, 1986).

47. See Michael Redclift, *Development and the Environmental Crisis: Red or Green Alternatives?* (London: Methuen, 1984) and Redclift, *Sustainable Development.*

48. Gorz, *Farewell to the Working Class*, 100.

49. Gorz, *Ecology as Politics,* chapter 1.

50. Enzensberger, "A Critique of Political Ecology," 21. Enzensberger's call has also been endorsed by Adrienne Farago, "Environmentalism and the Left," 13

51. See, for example, Lee, "On the Marxian View"; Herbert Marcuse, *One Dimensional Man* (London: Routledge and Kegan Paul, 1964; reprint ed., London: Abacus, 1972); Marcuse, *Counterrevolution and Revolt* (London: Allen Lane, 1972)—see in particular the chapter "Nature and Revolution," 60–128, especially 63–64; K. D. Shifferd, "Karl Marx and the Environment," *The Journal of Environmental Education* 3 (1972): 39–42; Gorz, *Ecology as Politics*; Gorz, *Farewell to the Working Class*; Thompson, "The Death of a Contradiction," 20; and Michael Lowy, "The Romantic and the Marxist Critique of Modern Civilization," *Theory and Society* 16 (1987): 891–904.

52. Ely, "Marxism and Green Politics in West Germany," 26.

53. It should be noted that Lee's case had already been advanced as early as 1972 by K. D. Shifferd in "Karl Marx and the Environment" and by Herbert Marcuse in his essay "Nature and Revolution" in *Counterrevolution and Revolt* (1972). However, Lee's argument in "On the Marxian View" is a more suitable focus for present purposes since it is both more recent and more developed.

54. Lee, "On the Marxian View," 11.

55. Ibid., 8.

56. Ibid., 15.

57. Ibid., 11.

58. Ibid., 16.

59. Marx, *The Economic and Philosophical Manuscripts of 1844*, 113.

60. Marx, *Capital,* vol. 3, 820. Against passages of this kind, Schmidt has noted passages from the *Grundrisse* (which represents an important bridge between the young and mature Marx) that suggest that Marx believed that a humanized realm of necessity can also become a sphere of human self-realization (*The Concept of Nature in Marx,* 143). This does not, however, detract from Marx's overriding concern to rationalize and reduce necessary labor. As Schmidt himself argues, "the problem of human freedom is reduced by Marx to the problem of *free time*" (142). This was also a concern of the young Marx (see Booth, "Gone Fishing") and certainly represents the direction in which Gorz and Marcuse have developed Marx's ideas.

61. According to Marcuse, Freud had theorized that "behind the reality principle lies the fundamental fact of Ananke or scarcity (*Lebensnot*), which means that the struggle for existence takes place in a world too poor for the satisfaction of human needs without constant restraint, renunciation, delay." See Herbert Marcuse, *Eros and Civilization: A Philosophical Inquiry into Freud* (London: Routledge and Kegan Paul, 1956), 35.

62. Balbus, *Marxism and Domination*, 274.

63. Benton, "Humanism = Speciesism," 7.

64. Balbus, *Marxism and Domination*, 274.

65. Routley, "On Karl Marx as an Environmental Hero," 239.

66. Ibid., 239–40.

67. Lee, "On the Marxian View," 9.

68. Benton, "Humanism = Speciesism," 8.

69. Ibid., 9.

70. There are passages in the *Grundrisse* that run contrary to this interpretation in that they suggest that human freedom or self-realization can be attained *through* democratic self-management in the workplace. This is a much more defensible interpretation from an ecocentric perspective. However, I argue that this interpretation is inconsistent with Marx's freedom/necessity distinction, which appears in both his early and mature writings.

71. Benton, "Humanism = Speciesism," 12.

72. Bookchin, "Marxism as Bourgeois Sociology," 204–6. See also Routley, "On Karl Marx as an Environmental Hero," 241.

73. Tolman, "Karl Marx, Alienation, and the Mastery of Nature," 72.

74. See also Benton, "Humanism = Speciesism," 14 and Routley, "On Karl Marx as an Environmental Hero," 242.

75. Marshall Sahlins, *Stone Age Economics* (London: Tavistock, 1974)—see especially chapter 1 "The Original Affluent Society," 1–39.

76. Ibid., 39.

77. Clark, "Marx's Inorganic Body," 250.

78. Isaac Balbus, "A Neo-Hegelian, Feminist, Psychoanalytical Perspective on Ecology," *Telos* 52: 140.

79. Hwa Yol Jung, "Marxism, Ecology, and Technology," *Environmental Ethics* 5 (1983): 170.

80. Booth, "Gone Fishing," 220.

81. As I pointed out in chapter 1, Mill and Bentham constitute two important exceptions within the liberal tradition insofar Mill entertained the idea of a stationary state economy and Bentham considered all sentient beings to be morally considerable.

NOTES TO CHAPTER 5

1. The Frankfurt School was founded in 1923 as an independently endowed institute for the exploration of social phenomenon. For a historical overview, see Martin Jay, *The Dialectical Imagination: A History of the Frankfurt School and the Institute of Social Research 1923–1970* (Boston: Little, Brown, 1973).

2. Friedrich Engels, "Socialism: Utopian and Scientific" in *The Marx-Engels Reader*, ed. Robert C. Tucker (New York: Norton, 1972), 638.

3. By "life-world" Habermas means "the taken-for-granted universe of daily social activity." Anthony Giddens, "Reason Without Revolution? Habermas's *Theories des kommunikativen Handelns,*" in *Habermas and Modernity*, ed. Richard J. Bernstein (Cambridge, U.K.: Polity, 1985), 101.

4. See, for example, Werner Hülsberg, *The German Greens: A Social And Political Profile* (London: Verso, 1988), 8–9 and John Ely, "Marxism and Green Politics in West Germany," *Thesis Eleven* no. 1 13 (1986): 27 and n. 11. It should be noted, however, that the themes of the *early* Frankfurt School theorists (Adorno, Horkheimer, and Marcuse) have had an important influence on the writings of Murray Bookchin, who has been an influential figure in the Green movement in North America. Bookchin was to *invert* the early Frankfurt School's thesis concerning the domination of human and nonhuman nature (see chapter 7).

5. For example, William Leiss, *The Domination of Nature* (Boston: Beacon, 1974); Timothy W. Luke and Stephen K. White, "Critical Theory, the Informational Revolution, and an Ecological Path to Modernity," in *Critical Theory and Public Life*, ed. John Forester (Cambridge: MIT Press, 1985), 22–53; and John Dryzek, *Rational Ecology: Environment and Political Economy* (Oxford: Blackwell, 1987).

6. See, for example, Peter Dews, ed., *Habermas: Autonomy and Solidarity* (London: Verso, 1986), 210.

7. Marcuse saw the ecology and feminist movements in particular as the most promising political movements and he foreshadowed many of the insights of ecofeminism. For example, in *Counterrevolution and Revolt* (London: Allen Lane, 1972), he argued for the elevation of the "female principle," describing the women's movement as a radical force that was undermining the sphere of aggressive needs, the performance principle, and the social institutions by which these are fostered (75).

8. Jürgen Habermas, "New Social Movements," *Telos* 49 (1981): 35. This article is extracted from the final chapter of Habermas, *The Theory of Communicative Action*, vol. 2: *Life-world and System: A Critique of Functionalist Reason*, trans. Thomas McCarthy (Boston: Beacon, 1987).

9. Habermas, "New Social Movements," 34.

10. Ibid., 35.

11. See Giddens, "Reason Without Revolution?" 121.

12. Murray Bookchin, "Finding the Subject: Notes on Whitebook and Habermas Ltd.," *Telos* 52 (1982): 83.

13. Theodor Adorno and Max Horkheimer, *Dialectic of Enlightenment*, trans. John Cummings (London: Verso, 1979)—this work was written during the second World War and first published in 1944. See also Herbert Marcuse, *Eros and Civilization: A Philosophical Inquiry into Freud* (London: Routledge and Kegan Paul, 1956) and *One Dimensional Man* (London: Routledge and Kegan Paul, 1964; reprint ed., London: Abacus, 1972).

14. Jay, *The Dialectical Imagination*, 256.

15. Ibid., 257 (see Adorno and Horkheimer, *Dialectic of Enlightenment*, 84 and 245–55). Frederick Engels' discussion of the subjugation of women in *The Origin of the Family, Private Property and the State* (London: Lawrence and Wishart, 1940) is, of course, an important exception.

16. See Martin Jay, "The Frankfurt School and the Genesis of Critical Theory," in *The Unknown Dimension: European Marxism Since Lenin*, ed. Dick Howard and Karl E. Klare (New York: Basic, 1972), 240–41.

17. Adorno and Horkheimer, *Dialectic of Enlightenment*, 9.

18. This theme has also been pursued by Eric Fromm in *Escape from Freedom* (New York: Holt, Rinehart, & Winston, 1969).

19. C. Fred Alford, *Science and the Revenge of Nature: Marcuse and Habermas* (Tampa/Gainesville: University Presses of Florida, 1985), 16.

20. Joel Whitebook, "The Problem of Nature in Habermas," *Telos* 40 (1979): 55.

21. Albrecht Wellmer, "Reason, Utopia, and the Dialectic of Enlightenment," *Praxis International* 3 (1983): 91.

22. Marcuse, *One Dimensional Man*.

23. Marcuse, *Counterrevolution and Revolt*, 74.

24. Ibid., 60.

25. Ibid. Marcuse argued that instead of seeing nature as mere utility, "the emancipated senses, in conjunction with a natural science proceeding on their basis, would guide the 'human appropriation' of nature" (ibid).

26. *One Dimensional Man*, 133–34.

27. Jürgen Habermas, *Toward a Rational Society: Student Protest, Science, and Politics*, trans. Jeremy J. Shapiro (London: Heinemann Educational Books, 1971), 85–87.

28. For example, Marcuse has stated: "The principles of modern science were *a priori* structured in such a way that they could serve as conceptual instruments for a universe of self-propelling, productive control; theoretical operationalism came to

correspond to practical operationalism. The scientific method [which] led to the ever-more-effective domination of nature thus came to provide the pure concepts as well as the instrumentalities for the ever-more-effective domination of man by man *through* the domination of nature" (*One Dimensional Man*, 130). And later: "The point which I am trying to make is that science, *by virtue of its own method* and concepts, has projected and promoted a universe in which the domination of nature has remained linked to the domination of man—a link which tends to be fatal to the universe as a whole" (ibid., 136).

29. William Leiss, "Technological Rationality: Marcuse and His Critics," *Philosophy of the Social Sciences* 2 (1972): 34–35. This essay also appears as an appendix to Leiss, *The Domination of Nature*, 199–212.

30. *One Dimensional Man*, 129.

31. *Ibid., 129 and 131.*

32. *Ibid., 137.*

33. *Marcuse*, Counterrevolution and Revolt, 61.

34. Alford, *Science and the Revenge of Nature*, 49–68.

35. Herbert Marcuse, *Eros and Civilization*, especially 35, 37, and 87–88.

36. Myriam Miedzian Malinovich, "On Herbert Marcuse and the Concept of Psychological Freedom," *Social Research* 49 (1982): 164.

37. Marcuse, *Counterrevolution and Revolt*, 64.

38. See Jürgen Habermas, "The Scientization of Politics and Public Opinion," in *Toward a Rational Society*, 62–80.

39. See, for example, Jürgen Habermas, "The Classical Doctrine of Politics in Relation to Social Philosophy," in *Theory and Practice*, trans. John Viertel (London: Heinemann, 1977), 41–81.

40. Jürgen Habermas, "Technology and Science as 'Ideology,'" in *Toward a Rational Society*, 107.

41. See Joel Whitebook, "The Problem of Nature in Habermas," 43.

42. Ibid.

43. Thomas McCarthy, *The Critical Theory of Jürgen Habermas* (Cambridge: Polity, 1984), 56.

44. Alford, *Science and the Revenge of Nature*, 77.

45. The norms that Habermas argues are implicit in every act of consensual communication are (i) that what each speaker says is intelligible or meaningful; (ii) that what each speaker says is true in terms of the propositional content of the statements; (iii) that each speaker communicates truthfully, with genuineness of intent

(i.e., without guile or dishonesty); and (iv) that what each speaker says is rationally justifiable.

46. Jürgen Habermas, *Knowledge and Human Interests*, trans. Jeremy Shapiro (London: Heinemann, 1972), 314.

47. Jürgen Habermas, *Legitimation Crisis*, trans. Thomas McCarthy (London: Heinemann, 1976); *The Theory of Communicative Action*, vol. 1: *Reason and the Rationalization of Society*, trans. Thomas McCarthy (Boston: Beacon, 1984); and *The Theory of Communicative Action*, vol. 2: *Life-world and System: A Critique of Functionalist Reason*, trans. Thomas McCarthy (Boston: Beacon, 1987).

48. For an overview of this general shift (in relation to vol. 1 of *The Theory of Communicative Action*), see Richard J. Bernstein's "Introduction" in Bernstein, ed., *Habermas and Modernity* (Cambridge, U.K.: Polity, 1985).

49. Bernstein, ed., *Habermas and Modernity*, 17.

50. Jürgen Habermas, "A Reply to My Critics," in *Habermas: Critical Debates*, ed. John B. Thompson and David Held (London: Macmillan, 1982), 238–50.

51. In *Legitimation Crisis* (41–43), for example, Habermas includes the ecology crisis as one of the many pathologies of modernity.

52. Dryzek, *Rational Ecology*, chapter 15.

53. Ibid., 204.

54. Ibid., 205.

55. Ibid., 206. Why is this so? Dryzek suggests, inter alia, that it "may be rooted in a recognition that to be accorded full subject status an entity must have the potential to participate in social discourse. Clearly, the entities of the natural world fail this test" (207).

56. Whitebook, "The Problem of Nature in Habermas," 61. It is conceivable that individuals in a communicatively rationalized society might collectively decide to extend their concept of ecological rationality beyond Dryzek's anthropocentric life-support approach by including other human interests in the nonhuman world (e.g., aesthetic, scientific, and recreational). The point, however, is that the framework would remain human centered.

57. Habermas, "A Reply to My Critics," 247.

58. Ibid., 248.

59. Giddens, "Reason Without Revolution?" 119.

60. Joel Whitebook, "The Problem of Nature in Habermas," 53.

61. On the moral considerability of animals, see, for example, Peter Singer, *Animal Liberation* (New York: Avon, 1975); on the moral considerability of plants and "natural objects," see, for example, Christopher D. Stone, *Should Trees Have Standing?: Toward Legal Rights for Natural Objects* (Los Altos, Calif.: Kaufmann, 1974).

62. John Clark, *The Anarchist Moment: Reflections on Culture, Nature and Power* (Montreal: Black Rose, 1984), 28.

63. Vincent Di Norcia, "From Critical Theory to Critical Ecology," *Telos* 22 (1974–75): 90 and 89.

64. See, for example, Henning Ottmann, "Cognitive Interests and Self-reflection," in *Habermas: Critical Debates,* ed. John B. Thompson and David Held (London: Macmillan, 1982), 78–97; McCarthy, *The Critical Theory of Jürgen Habermas,* 67; Alford, *Science and the Revenge of Nature,* chapter 9; Di Norcia, "From Critical Theory to Critical Ecology," 86–95; Whitebook, "The Problem of Nature in Habermas"; and Bookchin, "Finding the Subject: Notes on Whitebook and Habermas Ltd."

65. Henning Ottmann, "Cognitive Interests and Self-reflection," in *Habermas: Critical Debates,* ed. John B. Thompson and David Held, 89.

66. Ibid., 89.

67. Gregory Bateson, *Steps to an Ecology of Mind* (Frogmore, St. Albans: Paladin, 1973), 436–37.

68. For a pertinent discussion, see Andrew McLaughlin, "Is Science Successful? An Ecological View," *Philosophical Inquiry* 6 (1984): 39–46. See also McLaughlin, "Images and Ethics of Nature," *Environmental Ethics* 7 (1985): 293–319.

69. Di Norcia, "From Critical Theory to Critical Ecology," 95.

70. Jürgen Habermas, "A Reply to my Critics," 241.

71. Ibid., 245.

72. Alford, *Science and the Revenge of Nature,* 152–56.

73. Donald Worster, *Nature's Economy: A History of Ecological Ideas* (Cambridge: Cambridge University Press, 1985).

74. Alford, *Science and the Revenge of Nature,* 9–10. But Horkheimer and Adorno have also argued that "Science is not conscious of itself; it is only a tool. Enlightenment, however, is the philosophy which equates the truth with scientific systematization." (*Dialectic of Enlightenment,* 85).

75. Warwick Fox, *Toward a Transpersonal Ecology: Developing New Foundations for Environmentalism* (Boston: Shambhala, 1990), 253.

76. For a general discussion of how technology determines science, see Patsy Hallen, "What is Philosophy of Technology? An Introduction," *The Trumpeter* 5 (1988): 142–44.

77. Di Norcia, "From Critical Theory to Critical Ecology," 90.

78. See Mary Hesse, *Revolutions and Reconstructions in the Philosophy of Science* (Brighton: U.K.: Harvester, 1980), 187, and McLaughlin, "Images and Ethics of Nature," 295.

79. Evelyn Fox Keller, *A Feeling for the Organism: The Life and Work of Barbara McClintock* (New York: Freeman, 1983), and *Reflections on Gender and Science* (New Haven: Yale University Press, 1985).

80. Alford, *Science and the Revenge of Nature*, 140.

81. Whitebook, "The Problem of Nature in Habermas," 41.

82. See David Held, "Crisis Tendencies, Legitimation and the State," in *Habermas: Critical Debates*, ed. John B. Thompson and David Held (London: Macmillan, 1982), 187.

83. Luke and White, "Critical Theory, the Informational Revolution, and an Ecological Path to Modernity," 49.

84. Jürgen Habermas, "The Classical Doctrine of Politics in relation to Social Philosophy," in *Theory and Practice*, 41–81. Cf. John Rodman's discussion of the new ecological virtues of limits, community, and diversity in "Paradigm Change in Political Science," *American Behavioral Scientist* 24 (1980): 67–74.

NOTES TO CHAPTER 6

1. Raymond Williams, "Hesitations Before Socialism," *New Socialist*, September 1986, 35–36.

2. Frieder Otto Wolf, "Eco-Socialist Transition on the Threshold of the Twenty-First Century," *New Left Review* 158 (1986): 35.

3. Williams, "Hesitations Before Socialism," 34.

4. The slogan "neither left nor right" was coined by the West German C.D.U. dissident Herbert Gruhl, who failed to gain any support for his antisocialist platform in the critical debates that led up to *Die Grünen's* formation. See Werner Hülsberg, *The German Greens: A Social and Political Profile*, trans. Gus Fagan (London: Verso, 1988), 95–96. Commenting on the nature of ecopolitical debate in the United States, Daniel Faber and James O'Connor have observed that while Barry Commoner "has raised the issue of socialism, only in a few isolated circles has socialism been a central topic either philosophically or strategically." See Daniel Faber and James O'Connor, "The Struggle for Nature: Environmental Crisis and the Crisis of Environmentalism in the United States," *Capitalism, Nature, Socialism* 2 (1989): 33.

5. See, for example, Martin Ryle, *Ecology and Socialism* (London: Century Hutchinson, 1988), 91.

6. See, for example, *Die Grünen, Programme of the German Green Party* (London: Heretic Books, 1983).

7. See, for example, Stephen Bell, "Socialism and Ecology: Will Ever the Twain Meet?" *Social Alternatives* 6 (1987): 5–12.

8. David Pepper, "Radical Environmentalism and the Labour Movement," in

Red and Green: The New Politics of the Environment, ed. Joe Weston (London: Pluto, 1986), 116.

9. Joe Weston, "Introduction," in Pepper, *Red and Green: The New Politics of the Environment,* ed. Joe Weston (London: Pluto, 1986), 4–5.

10. Ibid., 5.

11. See Pepper, "Radical Environmentalism," 117.

12. André Gorz, *Ecology as Politics,* trans. Patsy Vigderman and Jonathan Cloud (London: Pluto Press, 1980), 3. Examples of ecosocialist theorists who conduct a similar analysis include Boris Frankel, *The Post-industrial Utopians* (Cambridge, U.K.: Polity in association with Basil Blackwell, 1987); Frankel, "Beyond Abstract Environmentalism," *Island Magazine,* Autumn 1989, 22–25; Bell, "Socialism and Ecology," 5–12; John Wiseman, "Red or Green? The German Ecological Movement," *Arena* 68 (1984): 38–56; and Richard Worthington, "Socialism and Ecology: An Overview," *New Political Science* 13 (1984): 69–83.

13. Garrett Hardin, "The Tragedy of the Commons," *Science* 162 (1968): 1243–48.

14. William E. Rees, "The Ecology of Sustainable Development," *The Ecologist* 20 (1990): 21.

15. Raymond Williams, *Towards 2000* (Harmondsworth, U.K.: Penguin, 1983), 214.

16. Ryle, *Ecology and Socialism,* 66.

17. André Gorz, *Farewell to the Working Class: An Essay on Post-Industrial Socialism* (London: Pluto, 1982), 100.

18. Williams, *Towards 2000,* 266.

19. Ibid., 262.

20. Gorz, *Farewell to the Working Class,* chapter 6.

21. André Gorz, *Paths to Paradise: On the Liberation from Work,* trans. Malcolm Imrie (London: Pluto, 1985), 35.

22. Ibid.

23. Gorz, *Farewell to the Working Class,* 75.

24. See, for example, Ryle, *Ecology and Socialism,* 88; Williams, *Towards 2000,* 254–55; and Carl Boggs, *Social Movements and Political Power: Emerging Forms of Radicalism in the West* (Philadelphia: Temple University Press, 1986), 19. According to Offe, new social movements tend to be supported by three main social groups or classes, namely, the New Class, decommodified groups [which roughly correspond to Gorz's "nonclass"], and disaffected members of the petty bourgeoisie. Claus Offe, "New Social Movements: Challenging the Boundaries of Institutional

Politics," *Social Research* 52 (1985): 832–38. Offe does point out, however, that unlike the petty bourgeoisie, *both* the "New Class" and "decommodified groups" are more likely to grow in number than disappear (837). For a critical discussion of explanations for the predominant New Class involvement in Green politics, see Robyn Eckersley, "Green Politics and the New Class: Selfishness or Virtue?" *Political Studies* 37 (1989): 205–23.

25. Ryle, *Ecology and Socialism,* 31 and 94.

26. James O'Connor, "Capitalism, Nature, Socialism: A Theoretical Introduction," *Capitalism, Nature, Socialism* 1 (1988): 37.

27. Williams, *Towards 2000,* 254.

28. Ibid., 254–55.

29. Ibid., 255.

30. Rudolf Bahro, *Socialism and Survival* (London: Heretic Books, 1982). Bahro's personal trajectory has moved from Marxism (*The Alternative in Eastern Europe* [London: New Left Books, 1978] to post-Marxist ecosocialism (*From Red to Green* [London: Verso/NLB, 1984] and *Socialism and Survival*) and finally to communal ecoanarchism (*Building the Green Movement* [London: Heretic/GMP, 1986]). *Socialism and Survival* may thus be seen as part of Bahro's "middle period." Some of his more recent ideas are discussed in chapter 7.

31. See, for example, Erik Dammann, *Revolution in the Affluent Society* (London: Heretic Books, 1984); F. E. Trainer, *Abandon Affluence!* (London: Zed, 1985); also by Trainer, *Developed to Death* (London: Merlin, 1989); and Frankel, *Post-industrial Utopians,* 261–62.

32. Williams, *Towards 2000,* 216. Williams also presents this center/periphery analysis in *The Country and the City* (London: Chatto and Windus, 1973).

33. Frankel, *Post-industrial Utopians,* 261–63.

34. Ryle, *Ecology and Socialism,* 70.

35. Ibid., 73–74.

36. Frankel, *Post-Industrial Utopians,* 260–63 and Ryle, *Ecology and Socialism,* 86.

37. Pepper, "Radical Environmentalism," 115. See also Pepper, *The Roots of Modern Environmentalism* (London: Croom Helm, 1984), 199.

38. See, for example, Ryle, *Ecology and Socialism,* 7–8; Weston, "Introduction," in Pepper, *Red and Green,* 2, and Pepper, "Radical Environmentalism," 121. See also Pepper, "Determinism, Idealism and the Politics of Environmentalism—A Viewpoint," *International Journal of Environmental Studies* 26 (1985): 11–19.

39. Indeed, Pepper describes such concerns as reactionary and "largely an elitist

defence of what a minority of ex-urbanites saw as 'wild nature' or 'traditional land-scapes.'" Pepper, "Radical Environmentalism," 121. See also Frankie Ashton, *Green Dreams, Red Realities*, N.A.T.T.A. Discussion Paper No. 2, Alternative Technology Group, The Open University, Milton Keynes, U.K., 1985.

40. See, for example, Weston, ed., *Red and Green* and Pepper, "Determinism, Idealism and the Politics of Environmentalism."

41. Gorz, *Ecology as Politics*, 16.

42. Ibid., 17.

43. David Pepper, *The Roots of Modern Environmentalism*, 193–94. See also Peter C. Gould, *Early Green Politics: Back to Nature, Back to the Land, and Socialism in Great Britain 1880–1900* (Brighton, U.K.: Harvester, 1988).

44. The terms of this debate were framed as early as 1974 by Hans Magnus Enzensberger in "A Critique of Political Ecology," *New Left Review* 84 (1974): 3–31. Since then, most of the discussion of the ecological crisis by ecosocialist theorists (e.g., Ryle, Gorz, Weston, Pepper, Bell, and Hülsberg) has been mainly couched in the language of "ecological limits" or "constraints" on human action.

45. Gorz, *Ecology as Politics*, 13.

46. For a discussion of the ecological ideas in Nazism, see Anna Bramwell, *Ecology in the 20th Century: A History* (Cambridge: Cambridge University Press, 1989).

47. Robert L. Heilbroner, *An Inquiry into the Human Prospect* (New York: Norton, 1974).

48. See, for example, Williams, *Towards 2000*, 214–15.

49. Gorz, *Ecology as Politics*, 18.

50. John Rodman, review of *Ecology as Politics*, by André Gorz, *Human Ecology* 12 (1984): 324.

51. Livingston, *The Fallacy of Wildlife Conservation* (Toronto: McClelland & Stewart, 1982), 42.

52. Raymond Williams, *Socialism and Ecology* (London: Socialist Environment and Resources Association, n.d.), 14. See also Williams, *The Country and the City*, 82.

53. See, for example, Enzensberger, "A Critique of Political Ecology," 13–15. Marx had argued that the apparent phenomenon of over-population under capitalism arose not as a result of natural conditions but rather as a result of the contradictions in the capitalist relations of production—in particular, its need to maintain an "industrial reserve army." For a discussion, see Michael Perelman, "Marx, Malthus, and the Concept of Natural Resource Scarcity," *Antipode* 11 (1979): 80–84.

54. These arguments received a considerable public airing in the debate

between Barry Commoner and Paul Ehrlich in the early 1970s. On the Ehrlich/Commoner debate, see Paul Ehrlich, *The Population Bomb,* revised ed., (London: Pan/Ballantine, 1972) and Barry Commoner, *The Closing Circle: Nature, Man and Technology* (New York: Bantam, 1972). For an exchange of views, see Paul Ehrlich, John Holdren, and Barry Commoner, "Dispute: *The Closing Circle,*" *Environment* 14 (1972): 24–25, 40–52.

55. See, for example, Pepper, *The Roots of Modern Environmentalism,* 167–69. On the more specific problem of world hunger, most ecosocialists focus on the need for land redistribution and a general shift in diet toward plant protein rather than on the need for birth control. See Frances Moore Lappé and Joseph Collins, with Cary Fowler, *Food First: Beyond the Myth of Scarcity* (New York: Ballantine, 1979). On the more general question of the human population explosion, see Frances Moore Lappé and Rachel Schuman, *Taking Population Seriously* (London: Earthscan Publications, 1989). Lappé and Schuman provide an excellent analysis of the power structures that contribute to high birth rates in developing countries. Although their primary focus is on the social causes and consequences of, and social solutions to, rapid population growth, Lappé and Schuman nonetheless argue (unlike most ecosocialists) that their analysis is capable of incorporating a nonanthropocentric perspective (see 70–71). In particular, they urge their readers to be cognizant of the impact of human population not only on humans but also on nonhuman life (4).

56. As Paul and Anne Ehlich point out, the key to understanding the role of human population growth in the environmental crisis lies in the equation $I = PAT$ (with I representing environmental impact, P representing the absolute size of the human population, A representing affluence or level of resource consumption, and T representing the environmental disruptiveness of the technologies that provide the resources consumed). See Paul R. Ehrlich and Anne H. Ehrlich, *The Population Explosion* (New York: Simon and Schuster, 1990), 58–59. This basic formula was first published in P. R. Ehrlich and J. P. Holdren, "Impact of Population Growth," *Science* 171 (1974): 1212–17.

57. For a brief discussion of this holistic model as it relates to socialism, see Worthington, "Socialism and Ecology," 78.

58. Williams, *Towards 2000,* 216.

59. Ibid., 256.

60. Ryle, *Ecology and Socialism,* 65.

61. Ibid., 60.

62. Ibid., 64.

63. Frankel, *Post-industrial Utopians,* 263.

64. See Wiseman et al., *New Economic Directions for Australia,* Discussion Paper, Department of Social Work, Phillip Institute of Technology, Melbourne, Victoria, (unpublished), 14.

65. Ibid., 14 and 48.

66. Ibid., 49–50.

67. Gorz, *Farewell to the Working Class*, 101.

68. See Gorz, *Farewell to the Working Class*, 96 and Ivan Illich, *Tools for Conviviality* (London: Calder and Boyars, 1973), 22–24.

69. Gore, *Farewell to the Working Class*, chapter 8.

70. Gore, *Ecology as Politics*, 15.

71. Ibid., 15–16.

72. Murray Bookchin, review of *Ecology as Politics*, by André Gorz, *Telos* 46 (1980–81): 182.

73. Richard Swift, "Liberation from Work," review of *Paths to Paradise*, by André Gorz, *Kick it Over*, Winter 1986–87, 17. Similarly, Boris Frankel has argued that Gorz "conceives of states in too narrow a political or administrative form" (*Post-industrial Utopians*, 63). In *Paths to Paradise* Gorz has modified his dual economy by introducing a third tier known as the "sphere of micro-social activity" that will partially mediate between the spheres of autonomy and heteronomy. This third tier would be "organized on a local level and based on voluntary participation, except where it replaces macro-social [i.e., heteronomous] work in providing for basic needs" (63). Such a modification does little to answer criticisms concerning Gorz's silence on the question of citizen participation in the sphere of socially necessary production.

74. See Frankel, *Post-industrial Utopians*, 93–97 and Alec Nove, *The Economics of Feasible Socialism* (London: George Allen & Unwin, 1983). For a general ecosocialist discussion of the kind of economic restructuring that might be relevant to the Australian economy, see Wiseman et al., *New Economic Directions for Australia*.

75. Many of the arguments that follow are drawn from Robyn Eckersley, "Ecosocialist Dilemmas: The Market Rules O.K.?," paper presented at the Socialist Scholars Conference, University of Sydney, 28 September–1 October 1990.

76. For a discussion of these issues, see James O'Connor, "Political Economy of Ecology of Socialism and Capitalism," *Capitalism, Nature, Socialism* 3 (1989): 93–107.

77. Ibid., 96.

78. "Introduction," *Alternatives to Capitalism*, ed. Jon Elster and Karl Ove Moene (Cambridge: Cambridge University Press, 1989), 4.

79. John Dryzek, *Rational Ecology: Environment and Political Economy* (Oxford: Blackwell, 1987), 106. See also Andrew McLaughlin, "Ecology, Capitalism, and Socialism," *Socialism and Democracy* 10 (1990): 92–93.

80. These criticisms are generally directed toward the long term feasibility of ecosocialism and do not address the more immediate (and difficult) problems of transition, such as capital flight.

81. Wiseman et al., *New Economic Directions for Australia*, 51.

82. Ibid.

83. Ryle, *Ecology and Socialism*, 42.

84. See, for example, the TOES anthology *The Living Economy: A New Economics in the Making*, ed. Paul Ekins (London: Routledge and Kegan Paul, 1986).

85. Philip Sutton, "Managing the Market to Achieve Ecologically Sustainable Development," 1991, MS, 7.

86. Herman E. Daly, "The Steady-State Economy: Alternative to Growthmania," MS, April, 1987, 7.

87. Ibid., 7.

88. Herman E. Daly and John B. Cobb, Jr. *For the Common Good: Redirecting the Economy Toward Community, the Environment, and a Sustainable Future* (Boston: Beacon, 1989), 48–49. See also Paul Ekins, "Sustainable Consumerism," The New Economics Foundation, London, MS, 1989, 11.

89. Paul Ekins, "Economy, Ecology, Society, Ethics: A Framework for Analysis," Paper presented at the Second Annual International Conference on Socio-Economics, George Washington University, March 1990, 12.

90. Karl Polanyi, "Our Obsolete Market Mentality," *The Ecologist* 4 (1974): 216. See also Polanyi, *The Great Transformation* (Boston: Beacon, 1967).

91. See, for example, Ward Morehouse, ed., *Building Sustainable Communities: Tools and Concepts for Self-Reliant Economic Change* (New York: Bootstrap, 1989).

92. Daly and Cobb, *For the Common Good*, 145–46.

93. These measures would also be complemented by a sustainable population policy.

94. For some original proposals as to how governments might review long-range impacts, see Lester Milbrath, *Envisioning a Sustainable Society: Learning Our Way Out* (Albany: State University of New York Press, 1989), chapter 14, especially 296–300.

95. See, for example, David Kemball-Cook, Mallen Baker, and Chris Mattingly, eds., *The Green Budget* (London: Green Print, 1991), 15.

96. For a general discussion, see David Pearce, *Blueprint for a Green Economy* (London: Earthscan Publications, 1989), chapter 7.

97. See, for example, Shann Turnbull, "Social Capitalism as the Road to Com-

munity Self-Management," in *Building Sustainable Communities: Tools and Concepts for Self-Reliant Economic Change,* ed. Ward Morehouse (New York: Bootstrap, 1989), 73–79 and "Re-inventing Corporations," *The Journal of Employee Ownership Law and Finance* 2 (1990): 109–36.

98. The major difference between social capitalism and market socialism, as I understand it, is that the latter is concerned to ensure that the *state* should have effective control of investment decisions (through state control of the banking system; the abolition of the stock market; nationalization of key industries; and worker democracy) whereas the former is concerned to ensure that the *local community* should have effective control of investment decisions.

99. See Ekins, "Sustainable Consumerism."

100. James Robertson, *Future Work: Jobs, Self-employment, and Leisure After the Industrial Age* (Aldershot, U.K.: Gower, 1985), x.

NOTES TO CHAPTER 7

1. I say "by and large" since there are some theorists in this current who are ex-Marxists and whose work bears the stamp of this legacy (e.g., Murray Bookchin and Rudolf Bahro) and others (e.g., William Morris) about whom the question as to whether they are indeed a Marxist, utopian socialist, or anarchist/libertarian is hotly debated.

2. Bookchin's publications are too numerous to list exhaustively here. His major books include *Our Synthetic Environment* (New York: Alfred A. Knopf, 1962), published under the pseudonym Lewis Herber; *Post-Scarcity Anarchism* (Berkeley: Ramparts, 1971); *Toward an Ecological Society* (Montreal: Black Rose, 1980); *The Ecology of Freedom* (Palo Alto, Calif.: Cheshire, 1982); *The Modern Crisis* (Philadelphia: New Society, 1986); *The Rise of Urbanization and the Decline of Citizenship* (San Francisco: Sierra Club Books, 1987); and *Remaking Society: Pathways to a Green Future* (Boston: South End, 1990). The major articles by Bookchin relevant to the present discussion are "Toward a Philosophy of Nature—The Bases for an Ecological Ethic," in *Deep Ecology,* ed. Michael Tobias (San Francisco: Avant, 1984), 213–35; "Freedom and Necessity in Nature: A Problem in Ecological Ethics," *Alternatives* 13 (1986): 29–38; and "Thinking Ecologically: A Dialectical Approach," *Our Generation* 18 (1987): 3–40. These three articles (some of which have been slightly revised) have been reprinted in *The Philosophy of Social Ecology: Essays on Dialectical Naturalism* (Montreal: Black Rose, 1990). See also John Clark, ed., *Renewing the Earth: The Promise of Social Ecology* (London: Green Print, 1990).

3. Bookchin, "Thinking Ecologically," 4.

4. Bookchin, *Ecology of Freedom,* 342.

5. Bookchin, *Post-Scarcity Anarchism,* 70.

6. For a thoroughgoing response to these criticisms by a leading deep/transpersonal ecology theorist, see Warwick Fox, "The Deep Ecology-Ecofeminism Debate and its Parallels," *Environmental Ethics* 11 (1989): 5–25.

7. Murray Bookchin, "On the Last Intellectuals," *Telos* 73 (1987): 182.

8. Bookchin, *Ecology of Freedom,* 340.

9. See, for example, Murray Bookchin, "Beyond Neo-Marxism," *Telos* 36 (1978): 5–28 and the chapter "On Neo-Marxism, Bureaucracy, and the Body Politic," in *Toward an Ecological Society,* 211–48. See also John Clark, *The Anarchist Moment: Reflections on Culture, Nature and Power* (Montreal: Black Rose, 1984).

10. Murray Bookchin, review of *Ecology as Politics,* by André Gorz, *Telos* 46 (1980–81): 179.

11. See, for example, *Ecology of Freedom,* 275.

12. See Bookchin, "Social Ecology Versus 'Deep Ecology': A Challenge to the Ecology Movement," *Green Perspectives: Newsletter of the Green Program Project,* Summer 1987, 1–23. This is by no means the first critique of deep ecology delivered by Bookchin, but it is the most polemical and has become the most notorious. For some responses and counter-responses to this critique, see Kirkpatrick Sale, "Deep Ecology and its Critics," *The Nation,* 14 May 1988, 670–75; for Bookchin's reply ("As if People Mattered") and Sale's counter-reply ("Sale Replies"), see *The Nation,* 10 October 1988, 294; George Sessions, "Ecocentrism and the Greens: Deep Ecology and the Environmental Task," *The Trumpeter* 5 (1988): 65–69; Bill Devall, "Deep Ecology and its Critics," *The Trumpeter* 5 (1988): 55–60; for letters and responses by Bill McCormick, Bill Devall, George Sessions, and Arne Naess, see *Green Synthesis,* September 1988, 3–5; for a reply, see Murray Bookchin, "A Reply to My Critics," *Green Synthesis,* December 1988, 5–7; Brian Tokar, "Exploring the New Ecologies: Social Ecology, Deep Ecology and the Future of Green Political Thought," *Alternatives* 15 (1988): 30–43; contributions by Bill Devall, Mike Kaulbars, Bill McCormick, and Brian Tokar in *Alternatives* 16 (1989): 49–54; and Arne Naess, "Finding Common Ground," *Green Synthesis,* March 1989, 9–10. More recently, however, Bookchin has declared his support for Dave Foreman and the Earth First! movement in North America and for wilderness protection. See Murray Bookchin and Dave Foreman, "Looking for Common Ground," in *Defending the Earth: A Dialogue Between Murray Bookchin and Dave Foreman,* Murray Bookchin and Dave Foreman (Boston: South End, 1991).

13. For an earlier statement of this thesis, see Bookchin, *Post-Scarcity Anarchism,* 63.

14. Bookchin, *Ecology of Freedom,* 350.

15. Bookchin, "Thinking Ecologically," 7–8, n. 1.

16. Ibid., 7, n. 1.

17. Bookchin, *Ecology of Freedom,* 352.

18. Ibid., 353.

19. Ibid., 36–37.

20. Ibid., 29.

21. Ibid., 36.

22. Bookchin employs the term *hierarchy* to mean a relationship and sensibility of obedience and command that arises from a ranking of some strata as "higher" and others "lower" (ibid, 4). He contrasts hierarchy with relationships that are characteristic of ecosystems, such as "unity-in-diversity," "differentiation," and "complementarity."

23. Bookchin, *Ecology of Freedom,* 6–7.

24. Fox, "The Deep Ecology-Ecofeminism Debate," 15.

25. In his reply to criticisms of his social hierarchy thesis (see Robyn Eckersley, "Divining Evolution: The Ecological Ethics of Murray Bookchin," *Environmental Ethics* 11 [1989]: 99–116, and Warwick Fox, "The Deep Ecology-Ecofeminism Debate,") Bookchin has insisted that he is only concerned with the *ideology* of dominating nature, notably, "the projection of the idea of social domination and control into nature—not with transient behavior patterns that come or go as a result of opportunistic, often historically short-lived circumstances" (Bookchin, "Recovering Evolution: A Reply to Eckersley and Fox," *Environmental Ethics* 12 [1990]: 262). However, in a recent summary of his ideas, Bookchin has emphasized that his social hierarchy thesis represents, inter alia, an historical statement "that the domination of human by human *preceded* the notion of dominating nature" (*Remaking Society,* 44).

26. Karl Wittfogel, *Oriental Despotism: A Comparative Study of Total Power* (New Haven: Yale University Press, 1957).

27. Donald Worster, "Water and the Flow of Power," *The Ecologist* 13 (1983): 168–74.

28. The significant turning point in this saga was the Australian grassroots environmental campaign to prevent the damming of the Franklin river. For an account of the impact of this campaign upon Tasmanian public affairs, see R. A. Herr and B. W. Davis, "The Tasmanian Parliament, Accountability and the Hydro-Electricity Commission: The Franklin River Controversy," in *Parliament and Bureaucracy, Parliamentary Scrutiny of Administration: Prospects and Problems in the 1980s,* ed. J. Nethercote (Sydney: Hale and Iremonger, 1982), 268–79.

29. Worster, "Water and the Flow of Power," 172.

30. Bookchin, *Ecology of Freedom,* 243.

31. See, for example, the chapters "Energy, 'Ecotechnocracy' and Ecology" and "The Concept of Ecotechnologies and Ecocommunities," in Bookchin, *Toward an Ecological Society.*

32. Lorna Salzman, "Politics as if Evolution Mattered: Some Thoughts on Deep and Social Ecology," Paper presented at the Ecopolitics IV Conference, University of Adelaide, South Australia, 21–24 September 1989, 15. Fox also argues precisely along these lines in "The Deep Ecology-Ecofeminism Debate."

33. George Woodcock, "Anarchism: A Historical Introduction," in *The Anarchist Reader,* ed. George Woodcock (London: Fontana, 1983), 27.

34. J. Hughes, "Beyond Bookchinism: A Left Green Response," *Socialist Review* 3 (1989): 107.

35. Bookchin, "Thinking Ecologically," 35.

36. That this is a misinterpretation should be clear from my explanation and defence of ecocentrism in chapters 2 and 3. For example, we have seen that the central concern of deep/transpersonal ecology theorists such as Arne Naess, Bill Devall, George Sessions, Warwick Fox, Alan Drengson, and Freya Mathews is to cultivate a sense of identification or empathy with all of nature (*of which humans are part*). This identification or empathy stems from the realization of our *interdependence* with other life-forms. This can hardly be interpreted as an approach that "reifies" nature and sets it *apart* from humanity.

37. Bookchin, "Thinking Ecologically," 20.

38. Bookchin, *Ecology of Freedom,* 343.

39. Salzman, "Politics as if Evolution Mattered," 15.

40. Bookchin, *Ecology of Freedom,* 363.

41. Ibid., 364.

42. Fritjof Capra, *The Turning Point: Science, Society and the Rising Culture* (London: Fontana, 1983; reprint ed., 1985), 311.

43. Ibid., 312. See also Richard Dawkins, *The Blind Watchmaker* (Harlow: Longman, 1986; reprint ed., London: Penguin, 1988).

44. Bookchin, *Ecology of Freedom,* 354.

45. See George Sessions, "Anthropocentrism and the Environmental Crisis," *Humboldt Journal of Social Relations* 2 (1974): 73 and Warwick Fox, *Toward a Transpersonal Ecology: Developing New Foundations for Environmentalism* (Boston: Shambhala, 1990) 13–16 and 253.

46. Bookchin, "Thinking Ecologically," 32.

47. For example, Anne and Paul Ehrlich have argued that "whatever remaining relatively undisturbed land exists that supports a biotic community of any significance should be set aside and fiercely defended against encroachment." See Anne H. Ehrlich and Paul R. Ehrlich, *Earth* (New York: Franklin Watts, 1987), 242. For a more general discussion, see George Sessions, "Ecocentrism and Global Ecosystem Protection," *Earth First!* 21 December 1989, 26–28. For Bookchin's recent statements in support of wilderness protection, see Bookchin and Foreman, "Looking for Common Ground," in *Defending the Earth.*

48. On the vexed subject of human population, Bookchin's contribution has mainly been one of warning of the dangers of fascist solutions than of advocating specific measures to control numan numbers. In particular, Bookchin's position has emerged largely as a response to, and critique of, certain blunt statements made by prominent members (such as Dave Foreman) of the U.S. environmental movement Earth First!, which Bookchin has wrongly taken as representative of the views of deep ecology theorists. For example, Foreman has remarked that it is better to leave Ethiopian children to starve than "save these half dead children who will never live a whole life. Their development will be stunted." See Dave Foreman, "A Spanner in the Woods," interviewed by Bill Devall, *Simply Living* 2, no. 12 (n.d.): 43. Yet, as Fox points out, "it is as unreasonable for Bookchin to condemn the body of ideas known as deep ecology on the basis that he does as it would be for a critic of Bookchin to condemn the body of ideas known as social ecology on the basis of whatever personal views happen to be put forward by individual activists who support any environmental organization that claims to draw on social ecology principles" (Fox, "The Deep Ecology-Ecofeminism Debate," 20 n. 38). Needless to say, deep/transpersonal ecology theorists have made it clear that "faced with the problem of hungry children, humanitarian action is a priority." In addition to Fox, see Arne Naess, "Sustainable Development and the Deep Long-Range Ecology Movement," *The Trumpeter* 5 (1988): 141. See also George Sessions, "Ecocentrism and the Greens," 65–66; and Kirkpatrick Sale, "Deep Ecology and its Critics," 675. As already noted, this debate has now moved on with the reconciliation between Bookchin and Foreman. For a detailed account, see Murray Bookchin and Dave Foreman, *Defending the Earth.*

49. See Nathan Keyfitz, "The Growing Human Population," *Scientific American,* September 1989, 71.

50. Arne Naess, *Ecology, Community and Lifestyle,* trans. David Rothenberg (Cambridge: Cambridge University Press, 1989), 29.

51. Naess, "Sustainable Development," 140. Elsewhere, Naess has provocatively asked: "Are cultural diversity, development of the sciences and arts, and of course basic human needs not served by, let us say, 100 million?" (Naess, *Ecology, Community and Lifestyle,* 141). This, of course, is merely an arbitrary figure and should not be taken literally. Naess's point is simply that there is no reason to believe that there will be less cultural diversity with a long term human population target that is considerably *lower* than the present 5.4 billion people.

52. Robert Nisbet, *The Social Philosophers: Community and Conflict in Western Thought* (London: Heinemann, 1974), 320.

53. Ibid., 320.

54. Ibid.

55. Ibid., 322.

56. Ibid., 324.

57. Lynn White, Jr., "The Historical Roots of Our Ecologic Crisis, *Science* 155 (1967): 1207.

58. Ibid., 326.

59. Kirkpatrick Sale, *Dwellers in the Land: The Bioregional Vision* (San Francisco: Sierra Club Books, 1985); E. F. Schumacher, *Small is Beautiful: Economics as if People Really Mattered* (London: Abacus, 1973); Rudolf Bahro, *Building the Green Movement* (London: Heretic/GMP, 1986); and Theodore Roszak, *Person/Planet: The Creative Disintegration of Industrial Society* (London: Victor Gollanz, 1979; reprint ed., London: Paladin, 1981).

60. Peter C. Gould, *Early Green Politics: Back to Nature, Back to the Land, and Socialism in Britain 1880–1900* (Brighton, U.K.: Harvester, 1988), 17. See also E. P. Thompson, *William Morris: Romantic to Revolutionary* (New York: Pantheon, 1976), 27–32.

61. Thompson, *William Morris,* 24. See William Morris, *News From Nowhere and Selected Writings and Designs* (London: Longmans, Green & Co., 1933; reprint ed., Harmondsworth, U.K.: Penguin, 1980). For a discussion of Morris's Marxist credentials, see the 1976 Postscript in Thompson, *William Morris,* 763–819. On the popularity of *News From Nowhere* in West Germany, see Elim Papadakis, *The Green Movement in West Germany* (London: Croom Helm, 1984), 54–55.

62. See Rudolf Bahro, *Building the Green Movement,* especially 86–98. Ecocommunalism is something Bahro has arrived at in his more recent work. In his earlier publications (for example, *Socialism and Survival* [London: Heretic/GMP, 1982] and *From Red to Green* [London: Verso/NLB, 1984]) Bahro's position was closer to ecosocialism than ecocommunalism. For a general discussion of the trajectory of Bahro's thought since he left East Germany, see Robyn Eckersley, "The Prophet of Green Fundamentalism," review of *Building the Green Movement,* by Rudolf Bahro, *The Ecologist* 17 (1987): 120–22.

63. Gilbert F. LaFreniere, "World Views and Environmental Ethics," *Environmental Review* 9 (1985): 319. See also Alan Drengson, "The Ecostery Foundation of North America (T.E.F.N.A.): Statement of Philosophy," *The Trumpeter* 7 (1990): 12–16.

64. Edward Goldsmith, "The Fall of the Roman Empire: A Social and Ecological Interpretation," *The Ecologist* 5 (1975): 196–206. This is reprinted in Goldsmith, *The Great U-Turn: De-industrializing Society* (Hartland, U.K.: Green Books, 1988), 3–29.

65. Edward Goldsmith, "De-industrializing Society," in *The Great U-Turn,* 206. Goldsmith's particular characterization of these "laws" is set forth in "The Way: An Ecological World-view," *The Ecologist* 18 (1988): 160–85.

66. There are also some paternalistic features of Goldsmith's particular solution that put him at odds with the ecoanarchist tradition and more in the company of the survivalist ecopolitical theorists discussed in chapter 1. Some of the nonlibertarian

measures recommended by Goldsmith include support for the use of advertising techniques to "convert" people to pursuing less wasteful patterns of consumption and a proposal to enrol unemployed people in a "Restoration Corps," and, after graduation from that, a Civil Militia along the Swiss model, which would enhance local patriotism. Goldsmith, "De-industrializing Society," in *The Great U-Turn*, 204 and 208.

67. Roszak, *Person/Planet*, 298–99.

68. Ibid., 301.

69. Ibid., 309.

70. Ibid., 306.

71. Ibid., 312.

72. André Gorz, *Farewell to the Working Class* (London: Pluto, 1982), 108–10.

73. Ibid., 111.

74. Ibid.

75. Peter Berg and Raymond F. Dasmann, "Reinhabiting California," in *Reinhabiting a Separate Country: A Bioregional Anthology of Northern California*, ed. Peter Berg (San Francisco: Planet Drum Foundation, 1978), 217–20. For a discussion of the possible origins of the term, see James J. Parsons, "On 'Bioregionalism' and 'Watershed' Consciousness," *The Professional Geographer* 37 (1985): 4.

76. Berg and Dasmann, "Reinhabiting California," 218.

77. Morris Berman, *The Reenchantment of the World* (Ithaca: Cornell University Press, 1981), 294.

78. Berg and Dasmann, "Reinhabiting California," 217–18.

79. See, for example, Peter Berg, "Devolving Beyond Global Monoculture," *CoEvolution Quarterly*, Winter 1981, 24–28; Gary Snyder, *Turtle Island* (New York: New Directions, 1969); Snyder, "Reinhabitation," in *The Old Ways* (San Francisco: City Lights, 1977), 57–66; Kirkpatrick Sale, *Dwellers in the Land: The Bioregional Vision* (San Francisco: Sierra Club books, 1985); Sale, "Bioregionalism—A New Way to Treat the Land," *The Ecologist* 14 (1984): 167–73; Ernest Callenbach, *Ecotopia* (London: Pluto, 1978); David Haenke, *Ecological Politics and Bioregionalism* (Drury, Mo.: New Life Farm, 1984); Jim Dodge, "Living by Life: Some Bioregional Theory and Practice," *The CoEvolution Quarterly*, Winter 1981, 6–12; Morris Berman, *The Reenchantment of the World*; and Brian Tokar, *The Green Alternative: Creating an Ecological Future* (San Pedro, Calif.: R & E Miles, 1987).

80. See, for example, Bill Devall and George Sessions, *Deep Ecology: Living as if Nature Mattered* (Layton, Utah: Gibbs M. Smith, 1985), 21–24, and Murray Bookchin, "A Letter of Support," in *North American Bioregional Congress Proceedings, May 21–25 1984* (Drury, Mo.: New Life Farm, 1984), 77–78.

81. Berman, *The Reenchantment of the World,* 294.

82. Peter Berg, "Growing a Bioregional Politics," *RAIN,* July–August 1985, 14.

83. Don Alexander, "Bioregionalism: Science or Sensibility?" *Environmental Ethics* 12 (1990): 162.

84. Peter Berg, "What is Bioregionalism?" *The Trumpeter* 8 (1991): 7.

85. Sale, *Dwellers in the Land,* 96. Sale also argues (108) that the bioregional emphasis on diversity is such that it does not ultimately matter what political forms are chosen within a particular bioregion—indeed, it is to be expected that not every bioregion will follow the American liberal tradition—provided they serve bioregional principles, namely, human scale, conservation and stability, self-sufficiency and cooperation, decentralization, and diversity. Nonetheless, he suggests that these bioregional principles would generally (though not always) impel the "polity in the direction of libertarian, noncoercive, open, and more or less democratic governance."

86. In a federation, political power is divided between the component states and a federal government under a federal constitution; moreover, the federal government can enact laws within its purview that apply directly to the citizens in the component states. See Roger Scruton, *A Dictionary of Political Thought* (London: Pan, 1982), 86 and 170.

87. Michael Helm, "Bioregional Planning," *RAIN,* October–November 1983, 23.

88. Alexander, "Bioregionalism," 167–69.

89. Ibid., 171.

90. Sale, *Dwellers in the Land,* 96.

91. For a discussion, see Alexander, "Bioregionalism," 167–69.

92. Woodcock, "Anarchism: A Historical Introduction," 17.

93. Ibid., 19.

94. Roszak, *Person/Planet,* 138

95. Koula Mellos's reading of Bookchin's ecoanarchism as a petit bourgeois form of radicalism based on solitary or "asocial individual self-sufficiency" seems to completely miss Bookchin's emphasis on symbiosis and community (see Koula Mellos, *Perspectives on Ecology: A Critical Essay* [London: Macmillan, 1988], chapter 4).

96. For an example of how community censure operates in traditional societies, see Harold Barclay, *People Without Government* (London: Kahn & Averill with Cienfuegos, 1982).

97. Bookchin, review of *Ecology as Politics,* 182.

98. Stephen Rainbow has also criticized what he calls the "soft" Green, ultra-democratic approach for naively assuming that local people will always choose to

attract ecologically sensitive industry. See Stephen Rainbow, "Eco-politics in Practice: Green Parties in New Zealand, Finland, and Sweden," Paper presented to the Ecopolitics IV Conference, University of Adelaide, South Australia, 21–24 September 1989, 21.

99. Robert Paehlke, *Bucolic Myths: Towards a More Urbanist Environmentalism* (Toronto: Center for Urban and Community Studies, University of Toronto, 1986). See also Paehlke, *Environmentalism and the Future of Progressive Politics* (New Haven: Yale University Press, 1989), 156–57 and 244–50.

100. Stephen Rainbow, "Eco-politics in Practice," 36.

101. For a provocative critique of the city, see Roszak, *Person/Planet*, chapter 9.

102. See Peter Berg, "The Bioregion and Ourselves II," *Fourth World News* 25 (1988): 8 and "A Metamorphosis for Cities: From Gray to Green," *The Trumpeter* 8 (1991): 9–12. See also Bookchin, *The Rise of Urbanization*, chapter 8.

103. See also Robyn Eckersley, "Green Politics: A Practice in Search of a Theory?" *Alternatives* 15 (1988): 60.

104. Bookchin, "Self-Management and the New Technology," in *Toward an Ecological Society*, 118–21.

105. Bookchin, *The Rise of Urbanization*, 282.

106. As Capra has shown, systems that are characterized by stratified order have greater flexibility and resilience than nonstratified systems in the face of perturbations. Capra, *The Turning Point*, 303–4.

NOTES TO CONCLUSION

1. Andrew McLaughlin, "Homo Faber or Homo Sapiens?" *The Trumpeter* 6 (1989): 21.

2. On the useful distinction between passive and active anthropocentrism, see Warwick Fox, *Toward a Transpersonal Ecology: Developing New Foundations for Environmentalism* (Boston: Shambhala, 1990), 21.

3. Alex Comfort, "Preface," in Harold Barclay, *People Without Government* (London: Kahn & Averill with Cienfuegos, 1982), 8.

4. I must hasten to add here that while all ecoanarchists are fundamentalists, not all ecosocialists are realists; however, it is a fair generalization to say that ecosocialists incline toward a more realistic or pragmatic strategy in that they are concerned to bring about *systemic* changes, which, in turn, require the capture of state power through the ballot box.

5. Ruth Levitas, "Marxism, Romanticism and Utopia: Ernst Bloch and William Morris," *Radical Philosophy* (Spring 1989): 33.

6. E. P. Thompson, *William Morris: Romantic to Revolutionary* (New York: Pantheon, 1976), 791. For a stimulating discussion of ecotopian literature, see George Sessions, "Ecophilosophy, Utopias, and Education," *Journal of Environmental Education* 15 (1983): 27–42.

Bibliography

Adorno, Theodor, and Horkheimer, Max. 1979. *Dialectic of Enlightenment*. Translated by John Cummings. London: Verso.

Alexander, Don. 1990. "Bioregionalism: Science or Sensibility?" *Environmental Ethics* 12:161–73.

Alford, C. Fred. 1985. *Science and the Revenge of Nature: Marcuse and Habermas*. Tampa/Gainesville: University Presses of Florida.

Ashton, Frankie. 1985. *Green Dreams, Red Realities*. N.A.T.T.A. Discussion Paper no. 2. Alternative Technology Group. Milton Keynes. U. K.: The Open University.

Babberger, Joan. 1974. "The Myth of Matriarchy: Why Men Rule in Primitive Societies." In *Woman, Culture, and Society*, 263–80. Edited by Michelle Zimbalist Rosaldo and Louise Lamphere. Stanford: Stanford University Press.

Bahro, Rudolf. 1978. *The Alternative in Eastern Europe*. London: New Left Books.

———. 1982. *Socialism and Survival*. London: Heretic Books.

———. 1983. "Socialism, Ecology and Utopia: An Interview." *History Workshop* 16:91–99.

———. 1984. *From Red to Green*. London: Verso/NLB.

———. 1986. *Building the Green Movement*. London: Heretic/GMP.

Balbus, Isaac D. 1982. *Marxism and Domination: A Neo-Hegelian, Feminist, Psychoanalytical Theory of Sexual, Political, and Technological Liberation*. Princeton, N.J.: Princeton University Press.

———. 1982. "A Neo-Hegelian, Feminist, Psychoanalytic Perspective on Ecology." *Telos* 52:140–55.

Barclay, Harold. 1982. *People Without Government*. London: Kahn & Averill with Cienfuegos.

Barnet, Richard J. 1980. *The Lean Years: Politics in the Age of Scarcity*. London: Abacus.

237

Barney, Gerald O., study director. 1982. *The Global 2000 Report to the President: Entering the Twenty-First Century*. Vol. 1. Harmondsworth, U. K.: Penguin.

Bateson, Gregory. 1973. *Steps to an Ecology of Mind*. Frogmore, St. Albans: Paladin.

Bell, Stephen. 1987. "Socialism and Ecology: Will Ever the Twain Meet?" *Social Alternatives* 6:5–12.

Benson, John. 1978. "Duty and the Beast." *Philosophy* 53:529–49.

Benton, Ted. 1988. "Humanism = Speciesism: Marx on Humans and Animals." *Radical Philosophy* Autumn:4–18.

Beresford, Melanie. 1971. "Doomsayers and Eco-nuts: A Critique of the Ecology Movement." *Politics* 12:98–106.

Berg, Peter. 1981. "Devolving Beyond Global Monoculture." *CoEvolution Quarterly*, Winter, 24–28.

———. 1985. "Growing a Bioregional Politics." *RAIN*, July–August, 14–16.

———. 1988. "The Bioregion and Ourselves II." *Fourth World News* 25:8.

———. 1991. "A Metamorphosis for Cities: From Gray to Green." *The Trumpeter* 8:9–12.

———. 1991. "What is Bioregionalism?" *The Trumpeter* 8:6–8.

Berg, Peter, and Dasmann, Raymond F. 1978. "Reinhabiting California." In *Reinhabiting a Separate Country: A Bioregional Anthology of Northern California*, 217–20. Edited by Peter Berg. San Francisco: Planet Drum Foundation.

Berman, Morris. 1981. *The Reenchantment of the World*. Ithaca: Cornell University Press.

Bernstein, Richard J., ed. 1985. *Habermas and Modernity*. Cambridge, U.K.: Polity.

Berry, Wendell. 1977. *The Unsettling of America*. New York: Avon.

Birch, Charles, and Cobb, John B., Jr. 1981. *The Liberation of Life: From the Cell to the Community*. Cambridge: Cambridge University Press.

Blackham, H. J. 1976. *Humanism*. New York: International Publishing Service.

Blake, Trevor. 1987. "Ecological Contradiction: The Grounding of Political Ecology." *Ecopolitics II Proceedings*, 76–83. Hobart: Centre for Environmental Studies, University of Tasmania.

Boggs, Carl. 1986. *Social Movements and Political Power: Emerging Forms of Radicalism in the West*. Philadelphia: Temple University Press.

Bookchin, Murray [Lewis Herber, pseud.]. 1962. *Our Synthetic Environment*. New York: Knopf.

Bookchin, Murray. 1971. *Post-Scarcity Anarchism*. Berkeley: Ramparts.

———. 1978. "Beyond Neo-Marxism." *Telos* 36:5–28.

———. 1980. "Marxism as Bourgeois Sociology." In *Toward an Ecological Society*, 193–210. Montreal: Black Rose.

———. 1980. "On Neo-Marxism, Bureaucracy and the Body Politic." In *Toward an Ecological Society*, 211–48. Montreal: Black Rose.

———. 1980. *Toward an Ecological Society*. Montreal: Black Rose Books.

———. 1980–81. Review of *Ecology as Politics*, by André Gorz. *Telos* 46:176–90.

———. 1982. *The Ecology of Freedom: The Emergence and Dissolution of Hierarchy*. Palo Alto, Calif.: Cheshire.

———. 1982. "Finding the Subject: Notes on Whitebook and Habermas Ltd." *Telos* 52:78–98.

———. 1984. "A Letter of Support." In *North American Bioregional Congress Proceedings, May 21–25 1984*, 77–78. Drury, Mo.: New Life Farm.

———. 1984. "Toward a Philosophy of Nature—The Bases for an Ecological Ethic." In *Deep Ecology*, 213–35. Edited by Michael Tobias. San Francisco: Avant.

———. 1986. "Freedom and Necessity in Nature: A Problem in Ecological Ethics." *Alternatives* 13:29–38.

———. 1986. *The Modern Crisis*. Philadelphia: New Society.

———. 1987. "On the Last Intellectuals." *Telos* 73:182–85.

———. 1987. *The Rise of Urbanization and the Decline of Citizenship*. San Francisco: Sierra Club Books.

———. 1987. "Social Ecology Versus 'Deep Ecology': A Challenge to the Ecology Movement," *Green Perspectives: Newsletter of the Green Program Project* Summer, 1–23.

———. 1987. "Thinking Ecologically: A Dialectical Approach." *Our Generation* 18:3–40.

———. 1988. "As if People Mattered." *The Nation*, 10 October, 294.

———. 1988. "A Reply to My Critics," *Green Synthesis*, December, 5–7.

———. 1990. *The Philosophy of Social Ecology: Essays on Dialectical Naturalism*. Montreal: Black Rose.

———. 1990. "Recovering Evolution: A Reply to Eckersley and Fox." *Environmental Ethics* 12:253–74.

———. 1990. *Remaking Society: Pathways to a Green Future*. Boston: South End Press.

Bookchin, Murray, and Foreman, Dave. 1991. *Defending the Earth: A Dialogue Between Murray Bookchin and Dave Foreman*. Boston: South End, 1991.

Booth, William James. 1989. "Gone Fishing: Making Sense of Marx's Concept of Communism." *Political Theory* 17:205–22.

Bramwell, Anna. 1989. *Ecology in the 20th Century: A History*. Cambridge: Cambridge University Press.

Brown, Harrison. 1954. *The Challenge of Man's Future*. New York: Viking.

Brown, Lester R., gen. ed. 1984 and annually thereafter. *State of the World 1984: A Worldwatch Institute Report on Progress Toward a Sustainable Society*. New York: W. W. Norton.

Caldecott, Leonie, and Leland, Stephanie, eds. 1983. *Reclaim the Earth: Women Speak Out for Life on Earth*. London: Women's Press.

Callenbach, Ernest. 1978. *Ecotopia*. London: Pluto Press.

Callicott, J. Baird. 1980. "Animal Liberation: A Triangular Affair." *Environmental Ethics* 2:311–38.

———. 1982. "Traditional American Indian and Western European Attitudes Toward Nature: An Overview." *Environmental Ethics* 4:293–318.

———. 1985. "Intrinsic Value, Quantum Theory, and Environmental Ethics." *Environmental Ethics* 7:257–75.

———. 1985. Review of *The Case for Animal Rights*, by Tom Regan. *Environmental Ethics* 7:365–75.

———. 1986. "The Metaphysical Implications of Ecology." *Environmental Ethics* 8:301–16.

———. 1987. "The Conceptual Foundations of the Land Ethic." In *Companion to A Sand County Almanac*, 186–217. Edited by J. Baird Callicott. Madison: University of Wisconsin Press.

———. 1989. "What's Wrong with the Case for Moral Pluralism." Paper presented at the Pacific Division Meeting of the American Philosophy Association, Berkeley, 23 March.

———. 1991. "The Wilderness Idea Revisited." MS.

Campbell, Joseph. 1972. *Myths to Live By*. New York: Viking; Reprint ed. New York: Bantam, 1973.

Capra, Fritjof. 1983. *The Turning Point: Science, Society, and the Rising Culture*. London: Fontana. Reprint. 1985.

Carson, Rachel. 1962. *Silent Spring*. New York: Fawcett Crest. Reprint ed. Harmondsworth, U. K.: Penguin, 1970.

Castoriadis, Cornelius. 1981. "From Ecology to Autonomy." *Thesis Eleven* no. 3:7–22.

Catton, William R., Jr., and Dunlap, Riley E. 1978. "Environmental Sociology: A New Paradigm." *American Sociologist* 13:41–49.

————. 1980. "A New Ecological Paradigm for Post-Exuberant Sociology." *American Behavioral Scientist* 24:15–47.

Cheney, Jim. 1987. "Eco-feminism and Deep Ecology." *Environmental Ethics* 9:115–45.

Chodorow, Nancy. 1978. *The Reproduction of Mothering.* Berkeley and Los Angeles: University of Calif. Press.

Clark, John. 1984. *The Anarchist Moment: Reflections on Culture, Nature and Power.* Montreal: Black Rose.

————. 1989. "Marx's Inorganic Body." *Environmental Ethics* 11:243–58.

————. ed. 1990. *Renewing the Earth: The Promise of Social Ecology.* London: Green Print.

Clark, S. L. R. 1975. *The Moral Status of Animals.* Oxford: Clarendon Press.

Clow, Michael. 1982. "Alienation from Nature: Marx and Environmental Politics." *Alternatives* 10:36–40.

Cole, H. S. D.; Freeman, C.; Jahoda, M.; and Pavitt, K. L. R. 1973. *Thinking About the Future: A Critique of the Limits to Growth.* London: Chatto and Windus.

Comfort, Alex. 1982. "Preface" to Harold Barclay, *People Without Government,* 7–9. London: Kahn & Averill with Cienfuegos Press.

Commoner, Barry. 1971. *The Closing Circle: Nature, Man and Technology.* New York: Knopf. Reprint. New York: Bantam, 1972.

Cox, Susan Jane Buck. 1985. "No Tragedy on the Commons." *Environmental Ethics* 7:49–61.

Daly, Herman E. 1987. "The Steady-State Economy: Alternative to Growthmania." MS.

Daly, Herman E., and Cobb, John B. Jr. 1989. *For the Common Good: Redirecting the Economy Toward Community, the Environment, and a Sustainable Future.* Boston: Beacon.

Dammann, Erik. 1984. *Revolution in the Affluent Society.* London: Heretic Books.

Dawkins, Richard. 1986. *The Blind Watchmaker.* Harlow: Longman. Reprint. London: Penguin, 1988.

de Beauvoir, Simone. 1978. *The Second Sex.* Translated and edited by H. M. Parshley. New York: Knopf. Reprint. Harmondsworth, U. K.: Penguin, 1982.

Devall, Bill. 1979. "Reformist Environmentalism." *Humboldt Journal of Social Relations* 6:129–57.

———. 1979. "Streams of Environmentalism." MS.

———. 1980. "The Deep Ecology Movement." *Natural Resources Journal* 20:299–322.

———. 1988. "Deep Ecology and its Critics." *The Trumpeter* 5:55–60.

———. 1988. Letter. *Green Synthesis,* September, 4.

———. 1988. *Simple in Means, Rich in Ends: Practicing Deep Ecology.* Layton, Utah: Gibbs M. Smith.

———. 1989. "Tokar is Wrong." *Alternatives* 16:49–50.

Devall, Bill, and Sessions, George. 1985. *Deep Ecology: Living as if Nature Mattered.* Layton, Utah: Gibbs M. Smith.

Dews, Peter, ed. 1986. *Habermas: Autonomy and Solidarity.* London: Verso.

Diamond, Irene, and Orenstein, Gloria Feman, eds. 1990. *Reweaving the World: The Emergence of Ecofeminism.* San Francisco: Sierra Club Books.

Di Norcia, Vincent. 1974–75. "From Critical Theory to Critical Ecology." *Telos* 22:86–95.

Dinnerstein, Dorothy. 1977. *The Mermaid and the Minotaur: Sexual Arrangements and Human Malaise.* New York: Harper and Row.

Dodge, Jim. 1981. "Living by Life: Some Bioregional Theory and Practice." *The CoEvolution Quarterly,* Winter, 6–12.

Drengson, Alan R. 1980. "Shifting Paradigms: From the Technocratic to the Person-Planetary." *Environmental Ethics* 3:221–40.

———. 1981. "Compassion and Transcendence of Duty and Inclination." *Philosophy Today* Spring:34–45.

———. 1986. "Developing Concepts of Environmental Relationships." *Philosophical Inquiry* 8:50–65.

———. 1989. *Beyond Environmental Crisis: From Technocrat to Planetary Person.* New York: Peter Lang.

———. 1989. "Protecting the Environment, Protecting Ourselves: Reflections on the Philosophical Dimension." In *Environmental Ethics,* 35–52. Vol. 2. Edited by R. Bradley and S. Duguid. Vancouver: Simon Fraser University.

———. 1990. "The Ecostery Foundation of North America (T.E.F.N.A.): Statement of Philosophy." *The Trumpeter* 7:12–16.

———. 1990. "Forests and Forestry Practices: A Philosophical Overview." *Forest Farm Journal* 2:2–3.

Dryzek, John. 1987. "Discursive Designs: Critical Theory and Political Institutions." *American Journal of Political Science* 31:656–79.

———. 1987. *Rational Ecology: Environment and Political Economy.* Oxford: Blackwell.

Dunlap, Riley E., and Morrison, Denton E. 1986. "Environmentalism and Elitism: A Conceptual and Empirical Analysis." *Environmental Management* 10:581–89.

Easlea, Brian. 1981. *Science and Sexual Oppression: Patriarchy's Confrontation with Woman and Nature.* London: Weidenfeld & Nicolson.

Easthope, Gary, and Holloway, Geoff. 1989. "Wilderness as the Sacred: The Franklin River Campaign." In *Environmental Politics in Australia and New Zealand,* 189–201. Edited by Peter Hay, Robyn Eckersley, and Geoff Holloway. Hobart: Centre for Environmental Studies, University of Tasmania.

Eckersley, Robyn. 1986. "The Environment Movement as Middle Class Elitism: A Critical Analysis." *Regional Journal of Social Issues* 18:24–36.

———. 1987. "The Prophet of Green Fundamentalism." Review of *Building the Green Movement,* by Rudolf Bahro. *The Ecologist* 17:120–22.

———. 1987. "The Road to Ecotopia?: Socialism Versus Environmentalism." *Island Magazine,* Spring, 18–25; reprinted in *The Ecologist* 18 (1988): 142–47; *The Trumpeter* 5 (1988): 60–64; and in *First Rights: A Decade of Island Magazine,* 50–60. Edited by Andrew Sant and Michael Denholm. Elwood, Victoria: Greenhouse, 1989.

———. 1988. "Green Politics: A Practice in Search of a Theory?" *Alternatives* 15:52–61.

———. 1989. "Divining Evolution: The Ecological Ethics of Murray Bookchin." *Environmental Ethics* 11:99–116.

———. 1989. "Green Politics and the New Class: Selfishness or Virtue?" *Political Studies* 37:205–23.

———. 1990. "Ecosocialist Dilemmas: The Market Rules O.K.?" Paper presented at the Socialist Scholars Conference, University of Sydney, 28 September–1 October.

———. 1992. "Environmental Theory and Practice in the Old and New Worlds: A Comparative Perspective." *Alternatives.* Forthcoming.

———. 1992. "Green Versus Ecosocialist Economic Programmes: The Market Rules O.K.?" *Political Studies,* Forthcoming.

Egerton, Frank N. 1978. "Changing Concepts of the Balance of Nature." *Quarterly Review of Biology* 48:322–50.

Ehrenfeld, David. 1981. *The Arrogance of Humanism.* Oxford: Oxford University Press.

Ehrlich, Paul. 1972. *The Population Bomb*. Rev. ed. London: Pan/Ballantine.

Ehrlich, Paul; Holdren, John; and Commoner, Barry. 1972. "Dispute: *The Closing Circle*." *Environment* 14:24.

Ehrlich, P. R., and Holdren, J. P. 1974. "Impact of Population Growth." *Science* 171:1212–17.

Ehrlich, Paul and Ehrlich, Anne. 1970. *Population, Resources, Environment*. San Francisco: Freeman.

———. 1987. *Earth*. New York: Franklin Watts.

———. 1990. *The Population Explosion*. New York: Simon and Schuster.

Ekins, Paul, ed. 1986. *The Living Economy: A New Economics in the Making*. London: Routledge and Kegan Paul.

———. 1989. "Sustainable Consumerism." MS. The New Economics Foundation, London.

———. 1990. "Economy, Ecology, Society, Ethics: A Framework for Analysis," Paper presented at the Second Annual International Conference on Socio-Economics, George Washington University, March.

Elster, Jon, and Moene, Karl Ove, eds. 1989. *Alternatives to Capitalism*. Cambridge: Cambridge University Press.

Elton, Charles. 1930. *Animal Ecology and Evolution*. Oxford: Oxford University Press.

Ely, John. 1986. "Marxism and Green Politics in West Germany." *Thesis Eleven* no. 13:22–38.

Engels, Frederick. 1940. *The Origin of the Family, Private Property and the State*. Translated by Alick West and Dona Torr. London: Lawrence and Wishart. Reprint ed., 1946.

———. 1971. *The Condition of the Working Class in England*. 2d ed. Translated and edited by W. O. Henderson and W. H. Chaloner. Oxford: Blackwell.

———. 1972. "Socialism: Utopian and Scientific." In *The Marx-Engels Reader*, 605–39. Edited by Robert C. Tucker. New York: Norton.

———. 1987. *Dialectics of Nature*. Translated by Clemens Dutt. In Karl Marx and Friedrich Engels, *Collected Works*, 313–588. Vol. 25. London: Lawrence & Wishart.

Enzensberger, Hans Magnus. 1974. "A Critique of Political Ecology." *New Left Review* 84:3–31.

Evernden, Neil. 1984. "The Environmentalist's Dilemma." In *The Paradox of Envi-*

ronmentalism, Symposium Proceedings, 7–17. Edited by Neil Evernden. Downsview, Ont.: Faculty of Environmental Studies, York University.

Faber, Daniel, and O'Connor, James. 1989. "The Struggle for Nature: Environmental Crisis and the Crisis of Environmentalism in the United States." *Capitalism, Nature, Socialism* 2:12–39.

Falk, Richard A. 1972. *This Endangered Planet: Prospects and Proposals for Human Survival*. New York: Vintage.

Farago, Adrienne. 1985. "Environmentalism and the Left." *Urban Policy and Research* 3:11–15.

Feenberg, Andrew. 1979. "Beyond the Politics of Survival." *Theory and Society* 7:319–61.

Foreman, Dave. n.d. "A Spanner in the Woods," Interviewed by Bill Devall. *Simply Living* 2(12): 40–43.

Fox, Stephen. 1981. *John Muir and His Legacy*. Boston: Little Brown.

Fox, Warwick. 1984. "Deep Ecology: A New Philosophy of Our Time?" *The Ecologist* 14:194–200.

———. 1985. "Towards a Deeper Ecology?" *Habitat Australia*, August, 26–28.

———. 1986. *Approaching Deep Ecology: A Response to Richard Sylvan's Critique of Deep Ecology*. Environmental Studies Occasional Paper no. 20. Hobart: Centre for Environmental Studies, University of Tasmania.

———. 1986. "Ways of Thinking Environmentally (and Some Brief Comments on their Implications for Acting Educationally)." In *Thinking Environmentally...Acting Educationally: Proceedings of the Fourth National Conference of the Australian Association of Environmental Education*, 21–29. Edited by J. Wilson, G. Di Chiro, and I. Robottom. Melbourne: Victorian Association for Environmental Education.

———. 1989. "The Deep Ecology-Ecofeminism Debate and its Parallels." *Environmental Ethics* 11:5–25.

———. 1989. "The Meanings of 'Deep Ecology'," *Island Magazine*, Autumn, 32–35; *The Trumpeter* 7 (1990):48–50.

———. 1990. *Toward a Transpersonal Ecology: Developing New Foundations for Environmentalism*. Boston: Shambhala.

———. 1991. "New Philosophical Directions in Environmental Decision-making." In *Theoretical Issues in Environmentalism: Essays from Australia*. Edited by P. R. Hay, R. Eckersley, and G. Holloway. Hobart: University of Tasmania. Forthcoming.

Frankel, Boris. 1987. *The Post-industrial Utopians*. Cambridge, U.K.: Polity in association with Basil Blackwell.

———. 1989. "Beyond Abstract Environmentalism." *Island Magazine,* Autumn, 22–25.

French, Marilyn. 1986. *Beyond Power: Women, Men and Morality.* London: Abacus.

Fromm, Eric. 1969. *Escape from Freedom.* New York: Holt, Rinehart, & Winston.

Garb, Yaakov Jerome. 1985. "The Use and Misuse of the Whole Earth Image." *Whole Earth Review,* March, 18–25.

Giddens, Anthony. 1985. "Reason Without Revolution? Habermas's *Theories des kommunikativen Handelns.*" In *Habermas and Modernity,* 95–123. Edited by Richard J. Bernstein. Cambridge: Polity.

Gilligan, Carol. 1982. *In a Different Voice.* Cambridge: Harvard University Press.

Glacken, Clarence J. 1965. "The Origins of the Conservation Philosophy." In *Readings in Resource Management and Conservation,* 158–63. Edited by Ian Burton and Robert W. Kates. Chicago: Chicago University Press.

Gleick, James. 1987. *Chaos: Making a New Science.* New York: Viking.

Godfrey-Smith, William. 1979. "The Value of Wilderness." *Environmental Ethics* 1:309–19.

Goldsmith, Edward. 1975. "The Fall of the Roman Empire: A Social and Ecological Interpretation." *The Ecologist* 5:196–206. Reprinted in Goldsmith, *The Great U-Turn: De-industrializing Society,* 3–29. Hartland, U.K.: Green Books, 1988.

———. 1988. *The Great U-Turn: De-Industrializing Society.* Hartland, U. K.: Green Books.

———. 1988. "The Way: An Ecological World-view." *The Ecologist* 18:160–85.

Goldsmith, Edward; Allen, Robert; Allaby, Michael; Davoll, John; and Lawrence, Sam. 1972. *Blueprint for Survival.* Boston: Houghton Mifflin. Reprint. Harmondsworth, U. K.: Penguin.

Gorz, André. 1980. *Ecology as Politics.* Translated by Patsy Vigderman and Jonathan Cloud. London: Pluto.

———. 1982. *Farewell to the Working Class: An Essay in Post-Industrial Socialism.* Translated by Michael Sonenscher. London: Pluto.

———. 1985. *Paths to Paradise: On the Liberation from Work.* Translated by Malcolm Imrie. London: Pluto.

Gould, Peter C. 1988. *Early Green Politics: Back to Nature, Back to the Land, and Socialism in Great Britain 1880–1900.* Brighton, U.K.: Harvester.

Gouldner, Alvin. 1973. *The Coming Crisis of Western Sociology.* London: Heinemann.

Gray, Elizabeth Dodson. 1981. *Green Paradise Lost*. Wellesley, Mass.: Roundtable.

Gribbon, John. 1979. *Future Worlds*. London: Abacus.

Griffin, Susan. 1978. *Woman and Nature: The Roaring Inside Her*. New York: Harper and Row.

Griscom, Joan L. 1981. "On Healing the Nature/History Split in Feminist Thought." *Heresies* 13:4–9.

Die Grünen, 1983. *Programme of the German Green Party*. London: Heretic Books.

Habermas, Jürgen. 1971. *Toward a Rational Society: Student Protest, Science, and Politics*. Translated by Jeremy Shapiro. London: Heinemann Educational Books.

———. 1972. *Knowledge and Human Interests*. Translated by Jeremy Shapiro. London: Heinemann.

———. 1976. *Legitimation Crisis*. Translated by Thomas McCarthy. London: Heinemann.

———. 1977. *Theory and Practice*. Translated by John Viertel. London: Heinemann.

———. 1981. "New Social Movements." *Telos* 49:33–37.

———. 1982. "A Reply to My Critics." In *Habermas: Critical Debates*, 219–83. Edited by John B. Thompson and David Held. London: Macmillan.

———. 1984. *The Theory of Communicative Action*. Vol. 1: *Reason and the Rationalization of Society*. Translated by Thomas McCarthy. Boston: Beacon.

———. 1987. *The Theory of Communicative Action*. Vol. 2: *Life-world and System: A Critique of Functionalist Reason*. Translated by Thomas McCarthy. Boston: Beacon.

Haenke, David. 1984. *Ecological Politics and Bioregionalism*. Drury, Mo.: New Life Farm.

Hallen, Patsy. 1987. "Making Peace with Nature: Why Ecology Needs Feminism." *The Trumpeter* 4:3–14.

———. 1988. "What is Philosophy of Technology? An Introduction." *The Trumpeter* 5:142–44.

Hanks, Patrick, ed. 1983. *Collins Dictionary of the English Language*. London: Collins.

Hardin, Garrett. 1968. "The Tragedy of the Commons." *Science* 162:1243–48; reprinted in *Environmental Ethics*, 242–52. Edited by K. S. Shrader-Frechette. Pacific Grove, Calif.: Boxwood, 1981.

———. 1972. *Exploring New Ethics for Survival: The Voyage of the Spaceship Beagle*. New York: Viking.

————. 1988. "Commons Failing." *New Scientist,* 22 October, 76.

Hay, P. R. 1988. "Ecological Values and Western Political Traditions: From Anarchism to Fascism." *Politics* (U. K.) 8:22–29.

Hay, P. R., and Haward, M. G. 1988. "Comparative Green Politics: Beyond the European Context?" *Political Studies* 36:433–48.

Hays, Samuel P. 1959. *Conservation and the Gospel of Efficiency.* Cambridge: Harvard University Press.

————. 1987. *Beauty, Health and Permanence: Environmental Politics in the United States, 1955–1985.* Cambridge: Cambridge University Press.

Heffernan, James D. 1982. "The Land Ethic: A Critical Appraisal." *Environmental Ethics* 4:235–47.

Heilbroner, Robert L. 1974. *An Inquiry into the Human Prospect.* New York: Norton.

Held, David. 1982. "Crisis Tendencies, Legitimation and the State." In *Habermas: Critical Debates,* 181–95. Edited by John B. Thompson and David Held. London: Macmillan.

Helm, Michael. 1983. "Bioregional Planning." *RAIN,* October–November, 22–23.

Herr, R. A., and Davis, B. W. 1982. "The Tasmanian Parliament, Accountability and the Hydro-Electricity Commission: The Franklin River Controversy." In *Parliament and Bureaucracy, Parliamentary Scrutiny of Administration: Prospects and Problems in the 1980s,* 268–79. Edited by J. Nethercote. Sydney: Hale and Iremonger.

Hesse, Mary. 1980. *Revolutions and Reconstructions in the Philosophy of Science.* Brighton, U.K.: Harvester.

Hill, Thomas E., Jr. 1983. "Ideals of Human Excellence and Preserving Natural Environments." *Environmental Ethics* 5:211–24.

Hirsch, Fred. 1976. *Social Limits to Growth.* Cambridge: Harvard University Press.

Holsworth, Robert. 1979. "Recycling Hobbes: The Limits to Political Ecology." *The Massachusetts Review* 20:9–40.

Hughes, J. 1989. "Beyond Bookchinism: A Left Green Response." *Socialist Review* 3:103–8.

Hughes, J. Donald. 1983. *American Indian Ecology.* El Paso, Tex.: Texas Western Press.

Hülsberg, Werner. 1988. *The German Greens: A Social and Political Profile.* London: Verso.

Illich, Ivan. 1973. *Tools for Conviviality.* London: Calder and Boyars.

Inglehart, Ronald. 1977. *The Silent Revolution: Changing Values and Political Styles Among Western Publics*. Princeton, N.J.: Princeton University Press.

———. 1981. "Post-Materialism in an Environment of Insecurity." *The American Political Science Review* 75:880–900.

Jay, Martin. 1972. "The Frankfurt School and the Genesis of Critical Theory." In *The Unknown Dimension: European Marxism Since Lenin*, 224–48. Edited by Dick Howard and Karl E. Klare. New York: Basic.

———. 1973. *The Dialectical Imagination: A History of the Frankfurt School and the Institute of Social Research 1923–1970*. Boston: Little, Brown.

Jung, Hwa Yol. 1983. "Marxism, Ecology, and Technology." *Environmental Ethics* 5:169–71.

Kahn, Herman; Brown, William; and Martel, Leon. 1976. *The Next 200 Years*. London: Associated Business Programmes. Reprint ed. London: Abacus, 1978.

Katsiaficas, George. 1987. *The Imagination of the New Left: A Global Analysis of 1968*. Boston, Mass.: South End.

Kaulbars, Mike. 1989. "Tokar is Wrong." *Alternatives* 16:50–52.

Keller, Evelyn Fox. 1983. *A Feeling for the Organism: The Life and Work of Barbara McClintock*. New York: Freeman.

———. 1985. *Reflections on Gender and Science*. New Haven: Yale University Press.

Kemball-Cook, David; Baker, Mallen; and Mattingly, Chris. eds. 1991. *The Green Budget*. London: Green Print.

Keyfitz, Nathan. 1989. "The Growing Human Population." *Scientific American*, September, 71–77.

Kheel, Marti. 1990. "Ecofeminism and Deep Ecology: Reflections on Identity and Difference." In *Reweaving the World*, 128–37. Edited by Irene Diamond and Gloria Feman Orenstein. San Francisco: Sierra Club Books.

King, Ynestra. 1983. "Toward an Ecological Feminism and a Feminist Ecology." In *Machina Ex Dea: Feminist Perspectives on Technology*, 118–29. Edited by Joan Rothschild. New York: Pergamon.

Kraft, Michael E. 1978. "Analyzing Scarcity: The Politics of Social Change." *Alternatives* Winter:30–33.

LaChapelle, Dolores. 1988. *Sacred Land, Sacred Sex: Rapture of the Deep*. Silverton, Colo.: Finn Hill Arts.

LaFreniere, Gilbert F. 1985. "World Views and Environmental Ethics." *Environmental Review* 9:307–22.

Lappé, Frances Moore, and Collins, Joseph (with Fowler, Cary). 1979. *Food First: Beyond the Myth of Scarcity*. New York: Ballantine.

Lappé, Frances Moore, and Callicott, J. Baird. 1987. "Marx Meets Muir: Toward a Synthesis of the Progressive Political and Ecological Visions," *Tikkun* 2/3:16–21.

Lappé, Frances Moore, and Schuman, Rachel. 1989. *Taking Population Seriously.* London: Earthscan Publications.

Lauber, Volkmar. 1978. "Ecology, Politics and Liberal Democracy." *Government and Opposition* 13:199–217.

Lee, Donald. 1980. "On the Marxian View of the Relationship between Man and Nature." *Environmental Ethics* 2:3–16.

Leeson, Susan M. 1979. "Philosophic Implications of the Ecological Crisis: The Authoritarian Challenge to Liberalism." *Polity* 11:303–18.

Leiss, William. 1972. "Technological Rationality: Marcuse and His Critics." *Philosophy of the Social Sciences* 2:31–42. Reprinted as an appendix to William Leiss, *The Domination of Nature,* 199–212. Boston: Beacon, 1974.

———. 1974. *The Domination of Nature.* Boston: Beacon.

———. 1978. *The Limits to Satisfaction: On Needs and Commodities.* London: Marion Boyars.

Leopold, Aldo. 1949. *A Sand County Almanac.* Oxford: Oxford University Press.

Levitas, Ruth. 1989. "Marxism, Romanticism and Utopia: Ernst Bloch and William Morris." *Radical Philosophy* Spring:27–36.

Livingston, John. 1981. *The Fallacy of Wildlife Conservation.* Toronto: McClelland and Stewart.

———. 1984. "The Dilemma of the Deep Ecologist." In *The Paradox of Environmentalism,* 61–72. Edited by Neil Evernden. Downsview, Ont.: Faculty of Environmental Studies, York University.

Lowy, Michael. 1987. "The Romantic and the Marxist Critique of Modern Civilization." *Theory and Society* 16:891–904.

Luke, Timothy W., and White, Stephen K. 1985. "Critical Theory, the Informational Revolution, and an Ecological Path to Modernity." In *Critical Theory and Public Life,* 22–53. Edited by John Forester. Cambridge: MIT Press.

Lukes, Steven. 1973. *Individualism.* New York: Harper and Row.

Malinovich, Myriam Miedzian. 1982. "On Herbert Marcuse and the Concept of Psychological Freedom." *Social Research* 49:158–80.

Marcuse, Herbert. 1956. *Eros and Civilization: A Philosophical Inquiry into Freud.* London: Routledge and Kegan Paul.

———. 1964. *One Dimensional Man.* London: Routledge and Kegan Paul. Reprint ed. London: Abacus, 1972.

———. 1972. *Counterrevolution and Revolt*. London: Allen Lane.

Marietta, Don E. 1984. "Environmentalism, Feminism, and the Future of American Society." *The Humanist* 44:15.

Marx, Karl. 1964. *The Economic and Philosophical Manuscripts of 1844*. Translated by Martin Milligan. Edited by Dirk J. Struik. New York: International Publishers.

———. 1970. *Capital: A Critique of Political Economy*. Edited by Friedrich Engels. Translated by Samuel Moore and Edward Aveling. Vol. 1: *Capitalist Production*. London: Lawrence & Wishart.

———. 1970. *Capital: A Critique of Political Economy*. Edited by Friedrich Engels. Translated by Samuel Moore and Edward Aveling. Vol. 3: *The Process of Capitalist Production as a Whole*. London: Lawrence & Wishart.

———. 1973. *Grundrisse: Foundations of the Critique of Political Economy*. Translated by Martin Nicholas. London: Allen Lane, New Left Review. Reprint. New York: Vintage.

Marx, Karl, and Engels, Frederick. 1976. *The German Ideology*. Translated by Clemens Dutt, W. Lough, and C. P. Magill. In Karl Marx and Frederick Engels, *Collected Works*, Vol. 5. London: Lawrence & Wishart, 1976.

Maturana, Humberto R., and Varela, Francisco J. 1988. *The Tree of Knowledge: The Biological Roots of Human Understanding*. Boston: Shambhala.

McCarthy, Thomas. 1984. *The Critical Theory of Jürgen Habermas*. Cambridge: Polity.

McConnell, Grant. 1965. "The Conservation Movement: Past and Present." In *Readings in Resource Management and Conservation*, 189–201. Edited by Ian Burton and Robert W. Kates. Chicago: Chicago University Press.

McConnell, Grant. 1971. "The Environmental Movement: Ambiguities and Meanings." *Natural Resources Journal* 11:427–35.

McCormick, Bill. 1988. Letter. *Green Synthesis*, September, 3.

———. 1989. "Tokar is Wrong." *Alternatives* 16:52–53.

McCormick, John. 1989. *The Global Environmental Movement: Reclaiming Paradise*. London: Belhaven.

McCulloch, Alistair. 1983. "The Ecology Party and Constituency Politics: The Anatomy of a Grassroots Party." Paper presented at the Annual Conference of the Political Studies Association, University of Newcastle-upon-Tyne, April.

McLaughlin, Andrew. 1984. "Is Science Successful? An Ecological View." *Philosophical Inquiry* 6:39–46.

———. 1985. "Images and Ethics of Nature." *Environmental Ethics* 7:293–319.

————. 1987. Review of *The Whale and the Reactor: A Search for Limits in an Age of High Technology*, by Langdon Winner. *Environmental Ethics* 9:377–80.

————. 1989. "Homo Faber or Homo Sapiens?" *The Trumpeter* 6:21–24.

————. 1990. "Ecology, Capitalism, and Socialism." *Socialism and Democracy* 10:69–102.

Meadows, Donella H.; Meadows, Dennis L.; Randers, Jorgen; and Behrens, William W., III. 1972. *The Limits to Growth: A Report for the Club of Rome's Project on the Predicament of Mankind*. New York: Universe.

Mellos, Koula. 1988. *Perspectives on Ecology: A Critical Essay*. London: Macmillan.

Merchant, Carolyn. 1982. *The Death of Nature: Women, Ecology and the Scientific Revolution*. London: Wildwood House.

Mesarovic, Mihajlo and Pestel, Eduard. 1974. *Mankind at the Turning Point*. New York: Dutton.

Midgley, Mary. 1983. *Animals and Why They Matter*. Harmondsworth, U. K.: Penguin.

Milbrath, Lester. 1984. *Environmentalists: Vanguard for a New Society*. Albany: State University of New York Press.

————. 1989. *Envisioning a Sustainable Society: Learning Our Way Out*. Albany: State University of New York Press.

Mill, J. S. 1979. *Principles of Political Economy*. Edited by Donald Winch. Harmondsworth, U. K.: Penguin.

Morehouse, Ward, ed. 1989. *Building Sustainable Communities: Tools and Concepts for Self-Reliant Economic Change*. New York: Bootstrap.

Morris, William. 1933. *News From Nowhere and Selected Writings and Designs*. London: Longmans, Green & Co. Reprint. Harmondsworth, U. K.: Penguin, 1933.

Mosley, J. G. 1972. "Toward a History of Conservation in Australia." In *Australia as Human Setting*, 136–54. Edited by Amos Rapaport. Sydney: Angus and Robertson.

Naess, Arne. 1973. "The Shallow and the Deep, Long-Range Ecology Movement. A Summary." *Inquiry* 16:95–100.

————. 1987. "Self-realization: An Ecological Approach to Being in the World." *The Trumpeter* 4:35–42.

————. 1988. Letter. *Green Synthesis*, September, 4–5.

————. 1988. "Sustainable Development and the Deep Long-Range Ecology Movement." *The Trumpeter* 5:138–42.

———. 1989. *Ecology, Community and Lifestyle.* Translated by David Rothenberg. Cambridge: Cambridge University Press.

———. 1989. "Finding Common Ground." *Green Synthesis,* March, 9–10.

Nash, Roderick. 1982. *Wilderness and the American Mind.* 3d ed. New Haven: Yale University Press.

———. 1985. "Rounding Out the American Revolution: Ethical Extensionism and the New Environmentalism." In *Deep Ecology,* 170–81. Edited by Michael Tobias. San Diego: Avant.

———. 1989. *The Rights of Nature: A History of Environmental Ethics.* Madison: University of Wisconsin Press.

Nisbet, Robert. 1974. *The Social Philosophers: Community and Conflict in Western Thought.* London: Heinemann.

Norton, Brian. 1986. "Sand Dollar Psychology." *The Washington Post Magazine,* 1 June, 10–14.

Noske, Barbara. 1989. *Humans and Other Animals.* London: Pluto.

Nove, Alec. 1983. *The Economics of Feasible Socialism.* London: George Allen & Unwin.

O'Connor, James. 1988. "Capitalism, Nature, Socialism: A Theoretical Introduction." *Capitalism, Nature, Socialism* 1:11–38.

———. 1989. "Introduction to Issue Number Two: Socialism and Ecology." *Capitalism, Nature, Socialism* 2:5–11.

———. 1989. "Political Economy of Ecology of Socialism and Capitalism." *Capitalism, Nature, Socialism* 3:93–107.

Offe, Claus. 1985. "New Social Movements: Challenging the Boundaries of Institutional Politics." *Social Research* 52: 832–38.

O'Riordan, Timothy. 1981. *Environmentalism.* 2d ed. London: Pion.

Ophuls, William. 1973. "Leviathan or Oblivion?" In *Toward a Steady State Economy,* 215–30. Edited by Herman E. Daly. San Francisco: Freeman.

———. 1974. "Reversal is the Law of Tao: The Immanent Resurrection of Political Philosophy." In *Environmental Politics,* 34–48. Edited by Stuart S. Nagel. New York: Praeger.

———. 1977. *Ecology and the Politics of Scarcity: A Prologue to a Political Theory of the Steady State.* San Francisco: Freeman.

Orr, David W., and Hill, Stuart. 1978. "Leviathan, the Open Society, and the Crisis of Ecology." *The Western Political Quarterly* 31:457–69.

Ottmann, Henning. 1982. "Cognitive Interests and Self-reflection." In *Habermas: Critical Debates*, 78–97. Edited by John B. Thompson and David Held. London: Macmillan.

Paehlke, Robert. 1986. *Bucolic Myths: Towards a More Urbanist Environmentalism.* Toronto: Centre for Urban and Community Studies, University of Toronto.

———. 1988. "Democracy, Bureaucracy, and Environmentalism." *Environmental Ethics* 10:291–308.

———. 1989. *Environmentalism and the Future of Progressive Politics.* New Haven: Yale University Press.

Papadakis, Elim. 1984. *The Green Movement in West Germany.* London: Croom Helm.

Parkin, Sara. 1989. *Green Parties: An International Guide.* London: Heretic.

Parsons, Howard. 1978. *Marx and Engels on Ecology.* Westport, Conn.: Greenwood.

Parsons, James J. 1985. "On 'Bioregionalism' and 'Watershed' Consciousness." *The Professional Geographer* 37:1–6.

Passmore, John. 1980. *Man's Responsibility for Nature: Ecological Problems and Western Traditions.* 2d ed. London: Duckworth.

Pearce, David. 1989. *Blueprint for a Green Economy.* London: Earthscan Publications.

Pepper, David. 1984. *The Roots of Modern Environmentalism.* London: Croom Helm.

———. 1985. "Determinism, Idealism and the Politics of Environmentalism—A Viewpoint." *International Journal of Environmental Studies* 26:11–19.

———. 1986. "Radical Environmentalism and the Labour Movement." In *Red and Green: The New Politics of the Environment*, 115–39. Edited by Joe Weston. London: Pluto.

Perelman, Michael. 1979. "Marx, Malthus, and the Concept of Natural Resource Scarcity." *Antipode* 11:80–84.

Pinchot, Gifford. 1910. *The Fight for Conservation.* New York: Doubleday Page and Co.

Plant, Judith, ed. 1989. *Healing the Wounds: The Promise of Ecofeminism.* Philadelphia: New Society.

Plumwood, Val. 1986. "Ecofeminism: An Overview and Discussion of Positions and Arguments." *Australasian Journal of Philosophy* 64:120–38.

———. 1988. "Women, Humanity and Nature." *Radical Philosophy* Spring:16–24.

Polanyi, Karl. 1967. *The Great Transformation.* Boston: Beacon.

————. 1974. "Our Obsolete Market Mentality." *The Ecologist* 4:213–20.

Porritt, Jonathon, and Winner, David. 1988. *The Coming of the Greens*. London: Fontana.

Po-Keung, Ip. 1983. "Taoism and the Foundations of Environmental Ethics." *Environmental Ethics* 5:335–43.

Pybus, Elizabeth M., and Broadie, Alexander. 1978. "Kant and the Maltreatment of Animals." *Philosophy* 53:560–61.

Rainbow, Stephen. 1989. "Eco-politics in Practice: Green Parties in New Zealand, Finland and Sweden." Paper presented at the Ecopolitics IV conference, University of Adelaide, South Australia, 21–24 September.

————. 1989. "New Zealand's Values Party: The Rise and Fall of the First National Green Party." In *Environmental Politics in Australia and New Zealand*, 175–88. Edited by Peter Hay, Robyn Eckersley, and Geoff Holloway. Hobart: University of Tasmania.

Rawls, John. 1976. *A Theory of Justice*. London: Oxford University Press.

Reader, John. 1988. "Human Ecology: How Land Shapes Society." *New Scientist*, 8 September, 51–55.

Redclift, Michael. 1984. *Development and the Environmental Crisis: Red or Green Alternatives?* London: Methuen.

————. 1985. "Marxism and the Environment: A View from the Periphery." In *Political Action and Social Identity: Class, Locality and Ideology*, 191–211. Edited by Gareth Rees, Janet Bujra, Paul Littlewood, Howard Newby, and Teresa L. Rees. London: Macmillan.

————. 1987. *Sustainable Development: Exploring the Contradictions*. London: Methuen.

Rees, William E. 1990. "The Ecology of Sustainable Development." *The Ecologist* 20:18–23.

Regan, Tom. 1980. "Animal Rights, Human Wrongs." *Environmental Ethics* 2:99–120.

————. 1983. *The Case for Animal Rights*. Berkeley and Los Angeles: University of Calif. Press.

Reich, Charles. 1971. *The Greening of America*. Harmondsworth, U. K.: Penguin.

Robertson, James. 1985. *Future Work: Jobs, Self-employment, and Leisure After the Industrial Age*. Aldershot, U. K.: Gower.

Rodman, John. 1973. "What is Living and What is Dead in the Political Philosophy of T. H. Green." *The Western Political Quarterly* 26:566–86.

———. 1977. "The Liberation of Nature?" *Inquiry* 20:83–145.

———. 1978. "Theory and Practice in the Environmental Movement: Notes Towards an Ecology of Experience." In *The Search for Absolute Values in a Changing World: Proceedings of the Sixth International Conference on the Unity of the Sciences*, 45–56. Vol 1. San Francisco: The International Cultural Foundation.

———. 1980. "Paradigm Change in Political Science: An Ecological Perspective." *American Behavioral Scientist* 24:49–78.

———. 1983. "Four Forms of Ecological Consciousness Reconsidered." In *Ethics and the Environment*, 82–92. Edited by Donald Scherer and Thomas Attig. Englewood Cliffs, N.J.: Prentice-Hall.

———. 1984. Review of *Ecology as Politics*, by André Gorz. *Human Ecology* 12:319–25.

Roszak, Theodore. 1970. *The Making of a Counterculture: Reflections on the Technocratic Society and its Youthful Opposition*. London: Faber & Faber. Reprint 1973.

———. 1973. *Where the Wasteland Ends: Politics and Transcendence in Postindustrial Society*. Garden City, N.Y.: Anchor/Doubleday. Reprint. London: Faber and Faber, 1973.

———. 1979. *Person/Planet: The Creative Disintegration of Industrial Society*. London: Victor Gollanz. Reprint. London: Paladin, 1981.

Rothschild, Joan, ed. 1983. *Machina Ex Dea: Feminist Perspectives on Technology*. New York: Pergamon.

Routley, R., and Routley V. 1979. "Against the Inevitability of Human Chauvinism." In *Ethics and the Problems of the 21st Century*, 36–59. Edited by K. E. Goodpaster and K. M. Sayre. Notre Dame: University of Notre Dame Press.

Routley, Richard, and Routley, Val. 1980. "Human Chauvinism and Environmental Ethics." In *Environmental Philosophy*, 96–189. Monograph Series, No. 2. Department of Philosophy, Research School of Social Sciences, Australian National University, Canberra, Australia.

Routley, Val. 1981. "On Karl Marx as an Environmental Hero." *Environmental Ethics* 3:237–44.

Ruether, Rosemary Radford. 1975. *New Woman New Earth: Sexist Ideologies and Human Liberation*. New York: Seabury.

Ryder, Richard. 1974. *Speciesism: The Ethics of Vivisection*. Edinburgh: Scottish Society for the Prevention of Vivisection.

Ryle, Martin. 1988. *Ecology and Socialism*. London: Century Hutchinson.

Sagoff, Mark. 1974. "On Preserving the Natural Environment." *The Yale Law Journal* 84:205–67.

————. 1988. *The Economy of the Earth*. Cambridge: Cambridge University Press.

Sahlins, Marshall. 1974. *Stone Age Economics*. London: Tavistock.

Sale, Kirkpatrick. 1984. "Bioregionalism—A New Way to Treat the Land." *The Ecologist* 14:167–73.

————. 1985. *Dwellers in the Land: The Bioregional Vision*. San Francisco: Sierra Club Books.

————. 1988. "Deep Ecology and its Critics." *The Nation*, 14 May, 670–75.

————. 1988. "Sale Replies." *The Nation*, 10 October, 294.

Salleh, Ariel Kay. 1984. "Deeper than Deep Ecology: The Ecofeminist Connection." *Environmental Ethics* 6:339–45.

Salzman, Lorna. 1989. "Politics as if Evolution Mattered: Some Thoughts on Deep and Social Ecology." Paper presented at the Ecopolitics IV Conference, University of Adelaide, South Australia, 21–24 September.

Sapontzis, Steve S. 1984. "Predation." *Ethics and Animals* 5:27–36.

Schmidt, Alfred. 1971. *The Concept of Nature in Marx*. London: New Left Books.

Schumacher, E. F. 1973. *Small is Beautiful: Economics as if People Really Mattered*. London: Abacus.

Scruton, Roger. 1983. *A Dictionary of Political Thought*. London: Pan.

Seed, John; Macy, Joanna; Fleming, Pat; and Naess, Arne. 1988. *Thinking like a Mountain: Towards a Council of All Beings*. Santa Cruz, Calif.: New Society.

Sessions, George. 1974. "Anthropocentrism and the Environmental Crisis." *Humboldt Journal of Social Relations* 2:71–81.

————. 1981. Review of *The Soul of the World: An Account of the Inwardness of Things*, by Conrad Bonifazi. *Environmental Ethics* 3:275–81.

————. 1983. "Ecophilosophy, Utopias, and Education." *Journal of Environmental Education* 15:27–42.

————. 1984. Review of *Eco-Philosophy: Designing New Tactics for Living*, by Henryk Skolimowski. *Environmental Ethics* 6:167–74.

————. 1987. "The Deep Ecology Movement: A Review." *Environmental Review* 11:105–25.

————. 1988. "Ecocentrism and the Greens: Deep Ecology and the Environmental Task." *The Trumpeter* 5:65–69.

————. 1988. Letter. *Green Synthesis*, September, 4.

————. 1989. "Ecocentrism and Global Ecosystem Protection." *Earth First!* 21 December, 26–28.

Sheldrake, Rupert. 1990. *The Rebirth of Nature: The Greening of Science and God.* London: Century.

Shepard, Paul. 1974. "Animal Rights and Human Rites." *North American Review,* Winter, 35–41.

Shifferd, K. D. 1972. "Karl Marx and the Environment." *The Journal of Environmental Education* 3:39–42.

Singer, Peter. 1975. *Animal Liberation: A New Ethics for Our Treatment of Animals.* New York: The New Review. Reprint. New York: Avon.

———. ed. 1985. *In Defence of Animals.* Oxford: Basil Blackwell.

Singleton, Fred. 1986. "Eastern Europe: Do the Greens Threaten the Reds?" *The World Today* 42:159–62.

Smith, Neil. 1984. *Uneven Development: Nature, Capital and the Production of Space.* Oxford: Blackwell.

Snyder, Gary. 1969. *Turtle Island.* New York: New Directions.

———. 1977. *The Old Ways.* San Francisco: City Lights.

———. 1984. "Anarchism, Buddhism, and Political Economy." Lecture delivered at the Fort Mason Centre, San Francisco, 27 February.

Spretnak, Charlene. N.d. *The Spiritual Dimension of Green Politics.* Santa Fe, N. Mex.: Bear and Company. Reprinted as Appendix C in Charlene Spretnak and Fritjof Capra, *Green Politics: The Global Promise,* 230–58. London: Paladin, 1986.

Spretnak, Charlene, and Capra, Fritjof. 1986. *Green Politics: The Global Promise.* London: Paladin.

Starhawk, 1982. *Dreaming the Dark: Magic, Sex, and Politics.* Boston: Beacon.

Stone, Christopher. 1974. *Should Trees Have Standing?: Toward Legal Rights for Natural Objects.* Los Altos, Calif.: Kaufmann.

Stretton, Hugh. 1976. *Capitalism, Socialism and the Environment.* Cambridge: Cambridge University Press.

Sutton, Philip. 1991. "Managing the Market to Achieve Ecologically Sustainable Development." MS, 1–48.

Swift, Richard. 1986–87. "Liberation from Work." Review of *Paths to Paradise,* by André Gorz. *Kick it Over,* Winter, 16–17.

Thompson, E. P. 1976. *William Morris: Romantic to Revolutionary.* New York: Pantheon.

Thompson, Janna. 1983. "The Death of a Contradiction: Marxism, the Environment and Social Change." *Intervention* 17:7–26.

Tokar, Brian. 1987. *The Green Alternative: Creating an Ecological Future.* San Pedro, Calif.: R. & E. Miles.

———. 1988. "Exploring the New Ecologies: Social Ecology, Deep Ecology and the Future of Green Political Thought." *Alternatives* 15:30–43.

———. 1989. "No, I'm Not [Wrong]." Reply to Devall, Kaulbars, and McCormick. *Alternatives* 16:53–54.

Tolman, Charles. 1981. "Karl Marx, Alienation, and the Mastery of Nature." *Environmental Ethics* 3:63–74.

Trainer, F. E. 1985. *Abandon Affluence!* London: Zed.

———. 1989. *Developed to Death.* London: Merlin.

Tribe, Laurence. 1974. "Ways Not to Think About Plastic Trees: New Foundations for Environmental Law." *The Yale Law Journal* 83:1315–48.

Turnbull, Shann. 1989. "Social Capitalism as the Road to Community Self-Management." In *Building Sustainable Communities: Tools and Concepts for Self-Reliant Economic Change,* 73–79. Edited by Ward Morehouse. New York: Bootstrap.

———. 1990. "Re-inventing Corporations." *The Journal of Employee Ownership Law and Finance.* 2:109–36.

Udall, Stuart L. 1963. *The Quiet Crisis.* New York: Holt, Rinehart, and Winston.

Varela, Francisco J.; Maturana, Humberto R.; and Uribe, Ricardo. 1974. "Autopoiesis: The Organization of Living Systems, Its Characterization and a Model." *Biosystems* 5:187–96.

Walker, K. J. 1988. "The Environmental Crisis: A Critique of Neo-Hobbesian Responses." *Polity* 21:67–81.

Walker, Pamela. 1989. "The United Tasmania Group: An Analysis of the World's First Green Party." In *Environmental Politics in Australia and New Zealand,* 161–74. Edited by Peter Hay, Robyn Eckersley, and Geoff Holloway. Hobart: University of Tasmania.

Warren, Karen J. 1987. "Feminism and Ecology: Making Connections." *Environmental Ethics* 9:3–20.

———. 1990. "The Power and the Promise of Ecological Feminism." *Environmental Ethics* 12:125–46.

Wellmer, Albrecht. 1983. "Reason, Utopia, and the Dialectic of Enlightenment." *Praxis International* 3:83–108.

Wells, David. 1978. "Radicalism, Conservatism and Environmentalism." *Politics* 13:299–306.

Wenz, Peter S. 1988. *Environmental Justice*. Albany: State University of New York Press.

Weston, Joe, ed. 1986. *Red and Green: The New Politics of the Environment*. London: Pluto.

White, Lynn, Jr. 1967. "The Historical Roots of Our Ecologic Crisis." *Science* 155:1203–7.

Whitebook, Joel. 1979. "The Problem of Nature in Habermas." *Telos* 40:41–69.

Williams, Raymond. 1973. *The Country and the City*. London: Chatto and Windus.

————. 1983. *Towards 2000*. Harmondsworth, U. K.: Penguin.

————. 1986. "Hesitations Before Socialism." *New Socialist,* September, 34–36.

————. n.d. *Socialism and Ecology*. London: Socialist Environment and Resources Association.

Winner, Langdon. 1986. *The Whale and the Reactor: A Search for Limits in an Age of High Technology*. Chicago: Chicago University Press.

Wiseman, John. 1984. "Red or Green? The German Ecological Movement." *Arena* 68:38–56.

Wiseman, John; Christoff, Peter; Watts, Rob; Read, Lorrie; Reid-Smith, Rob; Camilleri, Joe; Ward, Ian; and Frankel, Boris. 1988. *New Economic Directions for Australia*. Discussion Paper. Melbourne, Victoria: Phillip Institute of Technology.

Wittbecker, Alan E. 1986. "Deep Anthropology: Ecology and Human Order." *Environmental Ethics* 8:261–70.

Wittfogel, Karl. 1957. *Oriental Despotism: A Comparative Study of Total Power*. New Haven: Yale University Press.

Wolf, Frieder Otto. 1986. "Eco-Socialist Transition on the Threshold of the Twenty-First Century." *New Left Review* 158:32–42.

Woodcock, George. 1983. "Anarchism: A Historical Introduction." In *The Anarchist Reader*, 11–56. Edited by George Woodcock. London: Fontana.

World Commission on Environment and Development. 1987. *Our Common Future: The Report of the World Commission on Environment and Development*. Oxford: Oxford University Press.

Worster, Donald. 1983. "Water and the Flow of Power." *The Ecologist* 13:168–74.

————. 1985. *Nature's Economy: A History of Ecological Ideas*. Cambridge: Cambridge University Press.

Worthington, Richard. 1984. "Socialism and Ecology: An Overview." *New Political Science* 13:69–83.

Zimmerman, Michael. 1979. "Marx and Heidegger on the Technological Domination of Nature." *Philosophy Today* 23:99–112.

———. 1987. "Feminism, Deep Ecology, and Environmental Ethics." *Environmental Ethics* 9:21–44.

———. 1988. "Quantum Theory, Intrinsic Value, and Panentheism." *Environmental Ethics* 10:3–30.

Index

263

anthropocentrism, 127–28; defence of the state, 184; the dual economy of Andre Gorz, 134–36, 224n.73; ecological critique of capitalism, 120–23, 127, 132, ecological ethic of, 127–29; economic planning, 132–36; ecosocialist planning compared with Green market economy, 140; influence on Green movement, 119–20; the market versus planning debate, 136–40; passive anthropocentrism of, 180–81; relationship to ecocentric/anthropocentic dimension, 127–32, 179–81; relationship to Green movement, 125; relationship to Green political thought, 125; response to human population growth, 130, 223n.55; response to wilderness preservation, 130; role of the state in controlling market forces, 132–36; similarities with the Frankfurt School, 129; theoretical links with ecocentrism, 131–32

Ecosteries, 164

Ecotopian literature, 186

Eco-Marxism, 4, 145. *See also* Humanist eco-Marxism; Orthodox eco-Marxism

Ehrenfeld, David, 2, 56

Ehrlich, Anne, 223n.56, 229n.47

Ehrlich, Paul, 130, 222n.54, 223n.56, 229n.47

Ekins, Paul, 141

Elster, Jon, 137

Elton, Charles, 203n.45

Ely, John, 87

Emancipatory theory: anthropocentric/ecocentric cleavage within, 26–29; critique of industrialism, 22–23; new cultural emphasis in, 18–19; relationship to Critical Theory, 98; relationship to ecofeminism, 70–71; relationship to left/right dimension, 27; relationship to new social movements, 70–71. *See also* Green political thought

Engels, Frederick, 36, 68, 77, 79, 81–82, 84–85, 97, 215n.15

Enlightenment, 51, 86, 94, 101, 112, 114; idea of progress, 25; negative dialectics of, 98, 100, 102

Environmental cleavages, 26, 194n.73

Environmental crisis: characterization of, 7; official recognition of, 8

Environmental facism, 61

Environmental movement: birth of, 187n.2; growth of, 8; links with women's movement, 69, 214n.7; middle-class character of, 10. *See also* Environmentalism

Environmental philosophy, basic cleavage within, 26, 194n.73

Environmentalism: approaches to classification of, 33–35; major streams of, 33–35, 195n.1; relationship to political theory, 34, 193n. 60; general typologies of, 195n.3. *See also* Animal liberation; Ecocentrism; Human welfare ecology; Preservationism; Resource conservation

Enzensberger, Hans Magnus, 16, 87, 222n.44

Epistemologies of rule, 148, 151

Evernden, Neil, 36

Evolutionary ethics, 146; compared with ecocentrism, 155; of social ecology, 154–60; evolutionary stewardship thesis, 154–60

Evolutionary theory, relationship to social ecology, 156–57

Faber, Daniel, 219n.4

Fallacy of misplaced misanthropy, 202n.30

Farewell to the Working Class, 123

Fascism, 128–29. *See also* Environmental fascism

Federalism, 169, 233n.86. *See also* Confederalism

Feenberg, Andrew, 16

Feminism. *See* Ecofeminism

Feuerbach, Ludwig, 78, 209n.13

The Fight for Conservation, 35

Forces (means) of production, 79, 80–81, 84, 87, 116, 180